WORKBOOK
to accompany
MANAGING OUR
NATURAL RESOURCES
Fourth Edition

Delmar is proud to support FFA activities

Join us on the web at

www.Agriscience.Delmar.com

WORKBOOK
to accompany
MANAGING OUR NATURAL RESOURCES
Fourth Edition

Darold Hehn
William G. Camp
Thomas B. Daugherty

DELMAR
THOMSON LEARNING

Australia Canada Mexico Singapore Spain United Kingdom United States

COPYRIGHT © 2002, 2004 by Delmar,
a division of Thomson Learning, Inc. Thomson Learning™ is a trademark used herein under license

Printed in the United States of America
3 4 5 XXX 06 05 04 03

For more information contact Delmar,
3 Columbia Circle, PO Box 15015,
Albany, NY 12212-5015.

Or find us on the World Wide Web at http://www.delmar.com

ALL RIGHTS RESERVED. No part of this work covered by the copyright hereon may be reproduced or used in any form or by any means—graphic, electronic, or mechanical, including photocopying, recording, taping, Web distribution or information storage and retrieval systems—without written permission of the publisher.

For permission to use material from this text or product, contact us by
Tel   (800) 730-2214
Fax   (800) 730-2215
www.thomsonrights.com

Library of Congress Catalog Card Number: 00-064523
ISBN 0-7668-1556-0

Preface

This workbook supplements the use of the text *Managing Our Natural Resources, Fourth Edition*. The topics and activities are designed to provide hands-on instruction to students interested in the natural resources industry. The chapter worksheets provide the student an opportunity to outline the important textbook information for study and review.

Managing Our Natural Resources, Fourth Edition has established itself as the premier text for natural resources instruction in agriculture education classes. This text was published at a time when few natural resources curricula existed. *Managing Our Natural Resources, Fourth Edition* has successfully filled a void for natural resources instruction.

In the early 1970s some agriculture instructors began to teach natural resources to their agriculture students. Many instructors found that this type of instruction provided a good alternative to a strictly traditional production program. As programs moved away from these more traditional agriculture curricula, they were forced to give up some of the traditional hands-on and laboratory activities. More problems arose for these instructors as they began to realize that few hands-on natural resource curricula existed. This activities manual helps to fill this void with a wealth of hands-on and laboratory activities in the natural resources arena.

About the Author

Darold Hehn, the author of the activities manual, is the Agribusiness/Agriscience instructor at Rapid City Central High in Rapid City, South Dakota. He began teaching Natural Resources at the Huron Vocational School in Huron, South Dakota, in the 1970s. At that time Natural Resources was new to most agriculture curricula. At the time he was hired to the Rapid City School District, there were 15 students. He helped this natural resources based program grow into a four instructor department with over 500 students. He was the Outstanding Agriculture Teacher of the Year in South Dakota, the Agriscience Teacher of the Year in South Dakota, and received the State Vocational Director's Award for Innovation and Excellence.

**Dedicated to
Harold Hehn
his life was a blessing,
his death a sorrow.**

Contents

SECTION I   INTRODUCTION

Chapter 1   **Our Natural Resources** 2
    Test Your Knowledge   2
    Activity   4

Chapter 2   **A History of Conservation in the United States** 5
    Test Your Knowledge   5
    Activity   7

Chapter 3   **Concepts in Natural Resources Management** 8
    Test Your Knowledge   8
    Activity   10

Job Exercise for Chapters 1–3 11

SECTION II   SOIL AND LAND RESOURCES

Chapter 4   **Soil Characteristics** 14
    Test Your Knowledge   14
    Activity   16

Chapter 5   **Soil Erosion** 18
    Test Your Knowledge   18
    Activity   19

Chapter 6   **Controlling Erosion on the Farm** 21
    Test Your Knowledge   21
    Activity   22

Chapter 7   **Nonfarm Erosion Control** 24
    Test Your Knowledge   24
    Activity   25

Chapter 8   **Rangeland Management** 26
    Test Your Knowledge   26
    Activity   28

Chapter 9   **Landfills and Solid Waste Management** 30
    Test Your Knowledge   30
    Activity   32

Chapter 10   **Wetland Preservation and Management** 33
    Test Your Knowledge   33
    Activity   36

Chapter 11   **Land-Use Planning** 37
        Test Your Knowledge   37
        Activity   38

Chapter 12   **Careers in Soil Management** 39
        Test Your Knowledge   39
        Activity 1   40
        Activity 2   41

Job Exercise for Chapters 4–12 42

SECTION III   WATER RESOURCES

Chapter 13   **Water Supply and Water Users** 46
        Test Your Knowledge   46
        Activity   48

Chapter 14   **Water Pollution** 49
        Test Your Knowledge   49
        Activity   50

Chapter 15   **Water Purification and Wastewater Treatment** 52
        Test Your Knowledge   52
        Activity   54

Chapter 16   **Water-Use Planning** 56
        Test Your Knowledge   56
        Activity   57

Chapter 17   **Careers in Water Management** 59
        Test Your Knowledge   59
        Activity   60

Job Exercise for Chapters 13–17 61

SECTION IV   FOREST RESOURCES

Chapter 18   **Our Forests and Their Products** 64
        Test Your Knowledge   64
        Activity   67

Chapter 19   **Woodland Management** 69
        Test Your Knowledge   69
        Activity   71

Chapter 20   **Forest Enemies and Their Control** 73
        Test Your Knowledge   73
        Activity   74

Chapter 21   **Fire!** 76
        Test Your Knowledge   76
        Activity   78

Chapter 22   Careers in Forestry 79
          Test Your Knowledge   79
          Activity   80

Job Exercise for Chapters 18–22 81

SECTION V   FISH AND WILDLIFE RESOURCES

Chapter 23   Fish and Wildlife in America 84
          Test Your Knowledge   84
          Activity   85

Chapter 24   Game Management 86
          Test Your Knowledge   86
          Activity   88

Chapter 25   Marine Fisheries Management 89
          Test Your Knowledge   89
          Activity   93

Chapter 26   Freshwater Fishery Management 94
          Test Your Knowledge   94
          Activity   97

Chapter 27   Careers in Fish and Wildlife Management 98
          Test Your Knowledge   98
          Activity   100

Job Exercise for Chapters 23–27 101

SECTION VI   OUTDOOR RECREATION RESOURCES

Chapter 28   Recreation on Public Lands 108
          Test Your Knowledge   108
          Activity   110

Chapter 29   Outdoor Safety 111
          Test Your Knowledge   111
          Activity   112

Chapter 30   Careers in Outdoor Recreation 114
          Test Your Knowledge   114
          Activity   116

Job Exercise for Chapters 28–30 118

SECTION VII   ENERGY, MINERAL, AND METAL RESOURCES

Chapter 31   Fossil Fuel Management 120
          Test Your Knowledge   120
          Activity   123

Chapter 32   **Alternative Energy Sources Management**   **125**
         Test Your Knowledge   125
         Activity   127

Chapter 33   **Metals and Minerals** **128**
         Test Your Knowledge   128
         Activity   130

Chapter 34   **Careers in Energy, Mineral, and Metal Resources** **131**
         Test Your Knowledge   131
         Activity   133

Job Exercise for Chapters 31–34 .... **135**

SECTION VIII   ADVANCED CONCEPTS

Chapter 35   **Advanced Concepts in Natural Resources Management** **138**
         Test Your Knowledge   138
         Activity   139

Job Exercise for Chapter 35 .... **141**

SECTION I
Introduction

CHAPTER 1
Our Natural Resources

TEST YOUR KNOWLEDGE

Complete the following:

1. At one time the world's natural resources were considered endless, boundless, and **inexhaustible gifts**.
2. Natural resources are objects, materials, creatures, or energy found in nature that can be put to use by **humans**.
3. Things we consider as **resources** were not always considered natural resources.
4. The United States has a total land area of 3,675,545 square **miles**.
5. **Topsoil** is the uppermost layer of soil, where we get our food and natural fibers.
6. The soil's major enemy is **erosion**.
7. We have lost **1/3** of our topsoil to erosion.
8. A new problem is the **conversion** of agricultural land to urban expansion, industrialization, highway construction, and other use.
9. Land use **planning** establishes priorities for land use.
10. **70** percent of the earth's surface is covered by water.
11. Water is a natural resource, when it can be put to use by **humans**.
12. In order to make water more useful to man, we must **slow** it as it returns to the sea.
13. Water was an early source of **power**.
14. Water resource management also includes the **control** of excess runoff.
15. There is plenty of water, but it may not be where it is **needed**.
16. Fish and wildlife are **undomesticated** animals and plants.
17. Wildlife include both game and **nongame** animals.
18. About **48** vertebrate species have been exterminated in our nation.
19. About **228** vertebrate species are considered endangered in the United States.
20. Wildlife in this country is important, less for **food** and more for pleasure, like hunting and fishing.
21. Recently fish and wildlife management has begun to deal with **conservation** and **new** aspects.
22. The United States contains **738** million acres of forest.
23. **1-3rd** of the U.S. forest land is noncommercial and not usable for forest production.
24. A **mature** forest is dominated by mature, slow-growing trees.
25. With good **management**, we can cut trees and still have more than before.
26. Most of our energy comes directly or indirectly from the **sun**.

27. Energy sources include:
    a. coal
    b. wind
    c. water
    d. sun
    e. oil
28. Widely used minerals include:
    a. iron
    b. copper
    c. aluminum
    d. magnesium
    e. lead
    f. zinc
    g. tin
29. Recreational resources include:
    a. forests
    b. lasts
    c. breaches
    d. mountains
    e. parks
    f. game animals
    g. fish

CHAPTER 1
Our Natural Resources

ACTIVITY

Purpose:

List recreational activities in your area.

Research:

1. Obtain information about local outdoor recreation establishments.

2. Identify the outdoor recreation establishments of your area.

Procedure:

List the names of the establishments in their respective areas detailed here:

Forests

Lakes

Beaches

Mountains

Parks

Game Animals

Game Fish

Observations:

1. How many outdoor recreation activities exist in your area?

2. How far away from your home is the nearest recreational activity?

3. What outdoor recreation activity have you participated in recently?

CHAPTER 2
A History of Conservation in the United States

TEST YOUR KNOWLEDGE

Complete the following:

1. Our _____ greatness was built upon our forests, water, iron, coal, oil, and other natural resources.
2. Wise management of our natural resources is beginning to replace shortsighted _____.
3. When the settlers came, the colonies were covered largely by _____ forests.
4. The New York Sporting Club sought to promote restrictions against _____.
5. Because passenger pigeons were killed during nesting season, the passenger pigeon became _____.
6. The _____ Act made the interstate transportation of game taken against state law a federal crime.
7. In 1918 the Migratory Bird Treaty Act was passed to protect _____ waterfowl.
8. In _____ the Duck Stamp Act required waterfowl hunters to purchase a duck stamp.
9. Today every _____ operates its own fish and wildlife agency.
10. The earliest recorded shortage of timber occurred in _____ , about 5,000 years ago.
11. The Romans had to _____ wood from their conquered lands.
12. In 1626 Plymouth Colony passed America's first ordinance controlling the _____ of timber.
13. Several colonies passed laws against _____ of the forest.
14. The first conservation effort in the United States was for live oaks used in _____ building.
15. The _____ Forestry Association was organized in 1875 to promote timber culture and forestry.
16. The Division of Forestry was created in _____.
17. The great expansion of the national forests occurred during President _____ Roosevelt's term.
18. The Weeks Law gave the President authority to _____ forest lands.
19. During the Great Depression, the _____ _____ _____ did work in our national forests.
20. After World War II, the _____ industry expanded its need for timber.
21. Today forestry has replaced wasteful use of forests with intensive forest _____.
22. Our forefathers thought that land was without _____.
23. Since the 1600s, _____ of our topsoil has been lost to erosion.
24. In 1935 the _____ _____ Service was established.

25. Soil _____ districts were established to conserve local soil and water resources.
26. Today when the NRCS helps develop a plan for soil and water conservation, the Agricultural Stabilization and Conservation Service helps in _____ the practices.
27. In the early days, settlers built homes only where there was adequate _____.
28. When cities were built, water was used as a means of _____ disposal.
29. Early efforts of water conservation were for water used as _____, and as a by-product of forestry.
30. The federal government became involved with water management for _____ control.
31. The Flood Control Act of 1936 helped develop efforts in control of sediment and _____.
32. Concern for clean water supplies and wastewater treatment generated federal legislation in the _____ and _____.
33. Today's water problems also include the _____ of our water supply and lowered water tables.

CHAPTER 2
A History of Conservation in the United States

ACTIVITY

Purpose:
Evaluate state and federal agencies.

Research:
1. Locate a local telephone book.

2. Find the area detailing state and federal governments.

3. Determine the addresses for the following government agencies, which may be located in your area. The addresses may be found in the telephone book or via the Internet.

Procedure:
Detail the information under the respective heading below.

  State Game, Fish and Parks

  U.S. Fish and Wildlife Agency

  National Park Service

  U.S. Forest Service

  State Forestry Agency

  U.S. Army Corps of Engineers

  Extension Service

  U.S. Travel and Tourism

  Natural Resources Conservation Service

  Agriculture Stabilization and Conservation Service

Observations:
1. What state and federal agencies are located in your town?

2. Which agencies are housed in the same buildings?

3. Which agencies have you visited in the recent past?

CHAPTER 3
Concepts in Natural Resources Management

TEST YOUR KNOWLEDGE

Complete the following:

1. Many of our _____ _____ are not going to last forever.
2. We must _____ our natural resources for the future.
3. Natural resources that can last forever regardless of human activities are _____.
4. Water is a nonexhaustible resource, yet water supplies may be very _____.
5. Natural resources that can be replaced by human efforts are considered _____.
6. We use and produce more wood each year because forests are considered _____ resources.
7. _____ resources are those that cannot be replaced or reproduced.
8. We can conserve an _____ resource, but once it is gone it is gone forever.
9. _____ is a science that deals with the complex relationships among living things and their environment.
10. An _____ is any partially self-contained environment.
11. An _____ is a political activist with a special interest in some aspect of the environment.
12. _____ _____ refers to any change in the composition, shape, size or structure of the plants or animals in the ecosystem.
13. In _____, plants convert water and carbon dioxide into sugar.
14. _____ is a process that involves the breaking down of foods into their components along with the release of energy.
15. _____ is the process by which organic matter is reduced to organic compounds.
16. The replacement of one species by another in an ecosystem is _____ _____.
17. A _____ is the biotic subsystem in an extensive ecosystem.
18. _____ have an ability to make massive changes that can affect every living thing.
19. All living things make up part of a _____ _____, or food web.
20. _____ _____ refers to the ability of an ecosystem to provide food and shelter for a given population level.
21. Population _____ are affected by availability of food, water, shelter, diseases and parasites, and predators.
22. The world human population reached _____ billion in the year 1999.
23. We have no choice but to use our _____ _____ as fully and wisely as we can.
24. _____ means using nature to produce the maximum long-range benefit for people.

25. Some things that have value to people must be _____ because once it is gone, it is gone forever.
26. _____ use means that we plan management activities for natural resources to produce more than one benefit.
27. The conservation _____ of our nation should be the development and protection of a quality environment that serves both nature and people.

CHAPTER 3
Concepts in Natural Resources Management

ACTIVITY

Purpose:

Evaluate local natural resources.

Research:

Review and define the following terms:

Renewable

Nonrenewable

Exhaustible

Nonexhaustible

Procedure:

1. Define renewable resources.

2. Define exhaustible resources.

Observations:

1. Place an X inside the box that best explains the resources you researched and defined.

2. Complete this chart by adding local resources and placing an X inside the box that best explains them.

Resource	Nonexhaustible	Exhaustible	Nonrenewable	Renewable
Forest				
Water				
Wildlife				
Gold				
Grass				

Job Exercise for CHAPTERS 1–3

Purpose:
Interview a local natural resources employee.

Research:
1. Choose a natural resource position.

2. Research the duties and responsibilities of that position.

Procedure:
1. Write questions to ask in the interview. You may use any of the following:
    a. What are the responsibilities of a person in your position?
    b. What are the rewards of a person in your position?
    c. What are the opportunities for advancement?
    d. What is the salary range for this position?
    e. What are the employment prospects for this position?
    f. What education is required?

2. Interview an employee who works in that position.

Observations:
1. What education do you need in order to be employed in this position?

2. What job responsibility does this position require that you would enjoy doing?

3. What job responsibility does this position require that you would not enjoy doing?

SECTION II
Soil and Land Resources

CHAPTER 4
Soil Characteristics

TEST YOUR KNOWLEDGE

Complete the following:

1. The layer of natural materials on the earth's surface containing both organic and inorganic material and capable of supporting plant life is _____.
2. Soil is so fragile it can be _____ almost overnight.
3. Soil is so complex it cannot be _____ once destroyed.
4. Soil contains:
    a.
    b.
    c.
    d.
5. _____ _____ is made up of dead plant and animal materials in varying stages of decay.
6. The _____ of each main soil component vary from one soil type to another.
7. Soil is formed _____.
8. _____ _____ are materials underlying soil from which the soil was formed.
9. The five categories of soil parent material are:
    a.
    b.
    c.
    d.
    e.
10. _____ are chemically uniform, inorganic substances.
11. Rocks are _____ of minerals.
12. The general groups of rock are:
    a.
    b.
    c.
13. _____ rocks are formed by cooling of molten materials.
14. _____ rocks are formed by solidification of sediments.
15. _____ rocks are reformed igneous or sedimentary rocks due to heat and pressure.
16. _____ deposits were deposited by glaciers.
17. _____ deposits are wind-blown silt.
18. Alluvial deposits are waterborne _____ left by moving fresh water.
19. _____ deposits are dead vegetation that has built up enough to support plant life.

20. _____ and _____ can cause rocks to crack into smaller pieces.
21. Minerals in rock may be _____ and dissolve when exposed to moisture.
22. _____ frozen in cracks of rocks can break the rock into pieces.
23. _____ may be blown against a larger rock and cause it to weather.
24. _____ move rocks and grind them into pieces.
25. The major weathering forces are:
    a.
    b.
    c.
    d.
    e.
26. The proportion of organic matter in soil is _____ percent.
27. Organic matter consists of decaying _____ and _____ parts.
28. Original tissue is organic matter that is _____.
29. _____ is organic matter that is decomposed to where it is unrecognizable.
30. Organic matter serves many functions:
    a.
    b.
    c.
    d.
    e.
31. Most soils have three distinct layers or _____.
32. Each horizon has _____.
33. The horizons are called:
    a. A =
    b. B =
    c. C =
34. The _____ is the most productive part of the soil.
35. _____ is the angle of the soil surface expressed as a percentage of rise and fall.
36. Texture refers to the proportion of _____, _____, and _____.
37. The natural ability of the soil to allow water to flow through it is _____ _____.
38. _____ _____ refers to the likelihood that a field will receive flood damage.
39. The degree that the soil has been damaged is _____.
40. _____ refers to the depth of the topsoil or the subsoil.
41. Land capability classes categorize soil according to its _____ potential.
42. _____ _____ have been done by the NRCS to help improve land use planning.

CHAPTER 4
Soil Characteristics

ACTIVITY

Purpose:

Describe the soil formation process.

Research:

1. Review information in Chapter 4 of *Managing Our Natural Resources, Fourth Edition* on soil characteristics and formation.

2. Go to the library and find two additional sources.

Procedure:

You are a historian for an environmental magazine. The year is 2456. You are operating a time machine that allows you to see the events of the past. You are viewing an island that protruded above the ocean surface thousands of years before. As you view the time machine, you see that the island begins as a barren rock, and grows into a beautiful plant and animal rich environment.

- Island should be rich with plants and animals.

Observations:

Write a two-page report on the soil formation process of the island using the following components of the soil formation process.

time	parent material
minerals	rocks
glacial deposits	loess deposits
alluvial deposits	organic deposits
weathering	temperature changes
water action	plant roots
ice expansion	mechanical grinding
organic matter	soil profile
topsoil	subsoil
slope	erosion

CHAPTER 5
Soil Erosion

TEST YOUR KNOWLEDGE

Complete the following:

1. Natural soil erosion is known as _____ erosion.
2. As long as soil is covered by _____, it is protected from excess erosion.
3. When normal erosion is sped up by the clearing of natural vegetation, it is called _____ _____.
4. Water causes soil erosion by:
    a.
    b.
5. Run-off erosion causes:
    a.
    b.
    c.
6. Wind has two actions that cause erosion:
    a.
    b.
7. _____ erosion is the gradual and uniform removal of surface soil.
8. _____ erosion causes small streamlets to be cut into soil.
9. Although _____ erosion is a severe problem, it may not be as serious as rill or sheet erosion.
10. _____ cause soil removal and sand drifts.

CHAPTER 5
Soil Erosion

ACTIVITY

Purpose:
Use of the texture triangle.

Research:
The textural triangle shown here can be used to determine the textural class of a soil sample. In order to use the triangle, you must first know the percentage of sand, silt, and clay in your sample. The percentage of sand, silt, and clay can be calculated by mechanical analysis.

Once you know percentages, locate the percent sand along the bottom of the triangle, the percent clay along the left side of the triangle, and the percent silt along the right side of the triangle. The textural class is determined at the point where the clay, silt, and sand intersect. If all lines intersect on a division line between the classes, move toward the finer textured soils.

Soil Textures and Their Particles

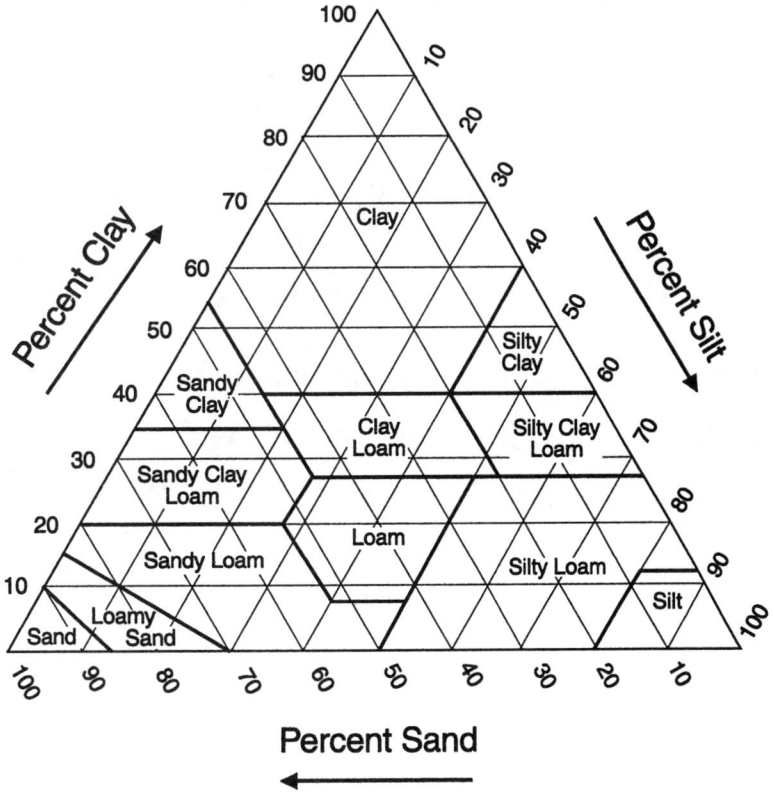

Procedure:

Given the relative amounts of sand, silt, and clay, find the textural classes of the following:

a. 25% sand _____
   20% clay
   55% silt

b. 50% sand _____
   40% clay
   10% silt

c. 70% sand _____
   25% clay
   5% silt

d. 40% sand _____
   42% clay
   18% silt

e. 10% sand _____
   80% clay
   10% silt

Observations:

1. What range of percent silt must a soil contain to be called a:

   a. Silt clay loam _____   c. Silty clay _____
   b. Silt loam   _____   d. Silt   _____

2. What range of percent sand must a soil contain to be called a:

   a. Sandy clay   _____   c. Sand   _____
   b. Sandy clay loam _____   d. Loamy sand _____

3. What range of percent clay must a soil contain to be called a:

   a. Clay loam _____   c. Sandy clay   _____
   b. Clay   _____   d. Sandy clay loam _____

CHAPTER 6
Controlling Erosion on the Farm

TEST YOUR KNOWLEDGE

Complete the following:

1. There are _____ land capability classes.
2. The use of land within its _____ is the most important step in land use.
3. There are two erosion-control categories:
   a.
   b.
4. On fields where erosion is severe, close-growing _____ _____ may provide protection.
5. A _____ _____ is an orderly repeated sequence of different crops grown on the same field.
6. _____ _____ is the production of alternating bands of different crops.
7. A _____ _____ is a drainageway permanently covered by vegetation.
8. Any tillage system that aids in soil and water conservation is called _____ _____.
9. Plowing around the hill is called _____ _____.
10. _____ accept the runoff and conduct it across the slope to some protected area.
11. _____ are rows of trees planted across the prevailing winds.
12. _____ are planted to provide protection for crops and livestock.

CHAPTER 6
Controlling Erosion on the Farm

ACTIVITY

Purpose:

Evaluate water/wind erosion control methods.

Research:

1. Review the erosion control methods for water and wind erosion.

2. Determine whether the controls are vegetative or mechanical.

Procedure:

Define the following erosion control methods and detail if they are vegetative controls (V) or mechanical controls (M).

crop rotation
_____
_____
_____
_____

cover crops
_____
_____
_____
_____

strip cropping
_____
_____
_____
_____

grassed waterways
_____
_____
_____
_____

conservation tillage
_____
_____
_____
_____

contour farming
_____
_____
_____
_____

terraces   windbreaks
_____   _____
_____   _____
_____   _____
_____   _____
_____   _____

shelterbelts
_____
_____
_____
_____
_____

Observations:
1. Which of the conservation methods are vegetative control?

2. Which of the conservation methods are mechanical control?

CHAPTER 7
Nonfarm Erosion Control

TEST YOUR KNOWLEDGE

Complete the following:

1. _____ involves more extreme changes on the soil than does farming.
2. Controlling nonfarm erosion involves:
   a.
   b.
   c.
3. A _____ ditch placed across the slope collects runoff and conducts it to safe disposal.
4. Concrete is used for construction of waterways because:
   a.
   b.
   c.
5. A very steep concrete lined waterway is called a water _____.
6. Sediment basins catch excess runoff long enough for _____ to settle out.
7. Banks are protected by:
   a.
   b.
   c.
   d.
   e.
8. Terraces break long _____ into shorter _____.
9. Lawns are vegetative control that is established by:
   a.
   b.
   c.
   d.
10. Mulches include:
    a.
    b.
    c.
    d.
    e.
    f.
11. _____ _____ such as shrubs and vines can be used to provide protection.
12. _____ _____ may provide a temporary cover crop.
13. Because of erosion problems, the law requires strip mines to have a _____ plan.

CHAPTER 7
Nonfarm Erosion Control

ACTIVITY

Purpose:

Calculate the topsoil needed.

Research:

The most common vegetative control technique for erosion control on nonfarm sites is the lawn. Sometimes topsoil must be applied before a lawn can be established. Topsoil is purchased by the cubic yard.

To determine the yards of topsoil needed, use the following formula:

$$\frac{\text{Length} \times \text{Width} \times (\text{Depth in inches})}{12} = \text{Cubic feet}$$

$$\frac{\text{Cubic feet}}{27} = \text{Cubic yards}$$

Example:

$$\frac{20 \times 10 \times 3}{12} = 50 \text{ Cubic feet}$$

$$\frac{50}{27} = 1.85 \text{ Cubic yards}$$

Procedure:

Calculate the topsoil needed for the following lawn areas:

1. A residential lawn measuring 100 feet long and 50 feet wide. The owner wants to add 4 inches of topsoil.

2. A commercial landscape with the following dimensions:

    a. Area #1: 50 feet long and 25 feet wide, 4 inches deep

    b. Area #2: 29 feet long and 8 feet wide, 2 inches deep

Observations:

1. What would be the cost for procedure #1 if topsoil cost $10.00 per yard?

2. What would be the cost for procedure #2 if topsoil cost $12.00 per yard?

CHAPTER 8
Rangeland Management

TEST YOUR KNOWLEDGE

Complete the following:

1. _____ refers to the land area that provides food, in the form of forage and browsing, for animals.
2. _____ refers to the land areas that tend to be naturally covered by grasses, grasslike plants, forbs, and shrubs.
3. Types of grassland in the United States include:
   a.
   b.
   c.
4. Because settlers recognized the _____ of the land and cleared it for farm production, the tallgrass prairies are almost gone today.
5. Native Americans used range _____ as a method of hunting for food.
6. Range fires killed _____ that would otherwise have become dominant.
7. _____ explorers released donkey, cattle, and other grazing animals to the grasslands.
8. The earliest _____ came to the grassland to look for gold, silver, or to harvest pelts.
9. Tall, abundant grass offered free grazing for _____.
10. The Taylor Grazing Act was designed to prevent _____ of grasslands.
11. Types of vegetation in rangelands include:
    a.
    b.
    c.
    d.
12. Value of grasses include:
    a.
    b.
13. Varieties of grass that are easily damaged by moderate grazing are called _____.
14. Rangeland plants that tend to thrive under heavy grazing are called _____.
15. Plants that move into an area after it has been badly overgrazed are called _____.
16. The primary objective of range management is the long-term _____ of livestock productivity from managed rangeland.

17. Specific objectives for rangeland management include:
    a.
    b.
    c.
    d.
    e.
    f.
18. The first step in planned range management is to determine the _____ _____ of the area.
19. Factors that affect carrying capacity are:
    a.
    b.
    c.
    d.
    e.
    f.
    g.
20. Grazing capacity is used to determine _____ rate.
21. An area's stocking rate is expressed in terms of animal _____ units.
22. One AEU is the amount of forage that is required to feed a _____ pound animal for a given period of time.
23. One _____ for a month is known as an animal unit month.
24. Complete the following:
    a. 1 steer   = _____ AEU
    b. 5 sheep/goats = _____ AEU
    c. 1 horse/bull   = _____ AEU
    d. 1 elk   = _____ AEU
    e. 4 deer   = _____ AEU
25. Once the grazing capacity of an area is determined, grazing rates must be _____.
26. Undergrazing and overgrazing discourage the growth of _____ plants.
27. Common management systems include:
    a.
    b.
    c.

CHAPTER 8
Rangeland Management

ACTIVITY

Purpose:

Categorize range plants.

Research:

1. Locate your state's Range and Pasture Judging Handbook or similar materials.

2. Locate the list of plants from your instructor.

3. Review the information about the following range plant characteristics:

   a. Grasses
   b. Sedges (grass-like)
   c. Forbs
   d. Shrubs

4. Review the following response characteristics:

   a. Desirable
   b. Undesirable
   c. Invader

Procedure:
1. Define "increaser."

2. Define "decreaser."

3. Define "invader."

Observations:
Complete this table by placing a check mark in the boxes that best explain the plant.

Plant	Grasses	Sedges	Forbs	Shrubs	Increaser	Decreaser	Invader

CHAPTER 9
Landfills and Solid Waste Management

TEST YOUR KNOWLEDGE

Complete the following:

1. _____ _____ management may be our most pressing remaining environmental problem.
2. Solid waste is _____ , non-soluble materials ranging from municipal garbage to industrial wastes.
3. _____ waste consists of spoilage from mining, logging, and other industrial processes, not disposed of in landfills.
4. The total amount of municipal solid waste has _____ in the last twenty years.
5. The three general types of solid waste are:
   a.
   b.
   c.
6. Hazardous wastes in sufficient quantities and concentrations pose a _____ to human life.
7. Wastes are considered hazardous if they contain any of the following characteristics:
   a.
   b.
   c.
   d.
   e.
8. _____ waste is any waste that contains, or is contaminated with, any radioactive material.
9. _____ waste is radioactive waste that also has the characteristics of a hazardous waste.
10. _____ _____ waste is solid waste that contains radioactive material or is contaminated with radioactive material.
11. Organic and aqueous liquids containing radioactive material is _____ waste.
12. Liquid Scintillation Vial waste is capped _____ vials containing either organic or biodegradable scintillation fluid contaminated with radioactive material.
13. Radioactive _____ are radioactively contaminated syringes, needles, surgical instruments, or other articles that have the potential to cut or puncture human skin.
14. Human societies have managed their solid waste for centuries by _____ it.
15. Water that enters a landfill and moves through the buried material is known as _____.

30

16. The leachate percolates downward, flowing out of the landfill as seepage or entering the _____.
17. Once contaminants are in the groundwater, they remain there _____ or until the water is removed for human use.
18. A _____ is an open area into which garbage is placed.
19. Landfills are covered by a layer of material, typically _____.
20. The two basic types of landfills are:
    a.
    b.
21. The natural attenuation landfill is designed to allow _____ percolation of precipitation to pass through the waste and through the underlying soil.
22. Paper products and yard waste are nonhazardous and can be _____ disposed of in a natural attenuation landfill.
23. List the three basic landfill shapes:
    a.
    b.
    c.
24. Waste is placed at or nearly at the normal soil level in an _____ landfill.
25. A _____ _____ landfill involves pushing the waste over the side of a hill or slope.
26. The most common landfill is the _____ _____ landfill.
27. A _____ is formed by removing the soil, the waste is placed in the trench, and the soil is used as the final cover, in a trench fill landfill.
28. _____ is the process by which the molecules of a chemical adhere to the surface of some other material.
29. Contaminants in leachate stick to _____ particles so tightly that they are permanently removed from the percolating water.
30. _____ removal takes place when bacteria, fungi, or other soil microorganisms break down or absorb the leachate constituents.
31. _____ _____ neutralizes the leachate by changing the molecular structure of the ions.
32. _____ means that the leachate is decreased by mixing it with large quantities of water.
33. _____ involves removal of solid constituents from the leachate by trapping them in pores in the soil.
34. When the liquid contaminant changes to a solid material, removing it from the leachate, it is called _____.
35. Total containment is necessary for certain hazardous and _____ waste.
36. A natural attenuation landfill can be made into a containment landfill by the addition of some sort of _____.
37. Recycling is important because of the _____ of solid waste disposal.

CHAPTER 9
Landfills and Solid Waste Management

ACTIVITY

Purpose:
  Identify landfill characteristics.

Research:
  1. Locate Internet Web sites that detail information about landfills. (www.enviroweb.issues/landfills/)

  2. Answer the following questions relating to landfills.

Procedure:
  1. What is a landfill?

  2. What are the elements of a landfill?

  3. What is a bottom liner?

  4. What are the different types of bottom liners?

  5. What is a cover?

  6. What are some problems with covers?

Observations:
  1. List the Internet Web site sources you used to answer the questions.

  2. Detail two "search" words you used to locate landfill information on the Internet.

CHAPTER 10
Wetland Preservation and Management

TEST YOUR KNOWLEDGE

Complete the following:

1. What is the ecological role of wetlands?
   a.
   b.
   c.
   d.
   e.
2. One-third of all threatened and endangered species live only in _____.
3. Plants that are attracted to water-saturated growing conditions are called _____.
4. _____ soils are soils that tend to be saturated with water most of the time.
5. A wetland is:
   a.
   b.
   c.
6. In our history, wetlands were considered a _____ rather than an asset.
7. List the two most common techniques for wetland identification:
   a.
   b.
8. Detail the three principal wetland resources available to landowners:
   a.
   b.
   c.
9. List the types of wetlands:
   a.
   b.
   c.
   d.
   e.
   f.
   g.
   h.

10. List the types of marshes:
    a.
    b.
    c.

11. List examples of marsh plants:
    a.
    b.
    c.
    d.
    e.

12. _____ are those areas that border rivers, lakes, and streams and that are flooded periodically.

13. List examples of wildlife found in ponds:
    a.
    b.
    c.
    d.
    e.
    f.
    g.

14. List the types of swamps:
    a.
    b.
    c.
    d.
    e.

15. _____ are areas that are very damp, usually with evergreens present, with a floor covered with moss and peat.

16. _____ _____ are usually full in the spring and early summer before water levels start to drop off.

17. _____ _____ last for only a few months each year.

18. There are over _____ million acres of wetlands in the United States.

19. Detail the major causes of the loss of wetlands:
    a.
    b.
    c.
    d.
    e.

20. The lower 48 states have lost over _____ of their original wetlands.

21. Wetland preservation efforts can be classified into these three areas:
    a.
    b.
    c.
22. List the causes of wetland loss:
    a.
    b.
    c.
    d.
    e.
    f.
    g.
    h.
    i.
    j.
    k.
23. Detail the part restoration plays in the management of degraded wetlands:
    a.
    b.
    c.
24. List the government programs that support wetlands:
    a.
    b.
    c.
    d.
    e.
    f.
    g.
    h.
    i.
    j.
    k.
    l.

CHAPTER 10
Wetland Preservation and Management

ACTIVITY

Purpose:

Evaluate a wetland habitat.

Research:

Review the information about wetlands detailed in the textbook.

Procedure:

Evaluate a local wetland found in your area.

Wetland Evaluation Data Form

Wetland name: _____ **Location:** _____

Category	Condition	
	Good	Poor
Water depth		
Vegetation species		
Wildlife species		
Land use of area		
Human-made alteration to site		
Contamination of site		
Asthetic value of site		
Water source for site		
Pollution controls at site		
Underground recharge ability		
Recreational opportunities		

Observations:

1. List the unique characteristics of the wetland.

2. What are the educational values of the wetland?

CHAPTER 11
Land-Use Planning

TEST YOUR KNOWLEDGE

Complete the following:

1. _____ planning refers to how the planning is done for an individual farm.
2. The governing body in each locality establishes the _____ and regulations concerning how the land can be used.
3. _____ regulations force like-activities to be grouped together.
4. The farmer's responsibilities include:
   a.
   b.
   c.
   d.
5. Farming operations must produce income to:
   a.
   b.
   c.
   d.
   e.
   f.
   g.
6. The United States has a total of _____ _____ acres of land.
7. Cities and towns make up _____ million acres.
8. Each year _____ million acres of prime farmland and _____ million acres of lesser-quality farmland are taken for nonfarm use.
9. _____ and _____ land use is more profitable than farm use, so a developer can always pay more than a farmer for land.
10. Land naturally tends to be put to its most _____ use.
11. Once an area is _____, it becomes lost for food production.

CHAPTER 11
Land-Use Planning

ACTIVITY

Purpose:

Identify symbols used on soil conservation maps.

Research:

1. Locate Figure 11-1 in the text.

2. Review the symbols used on a soil conservation map.

Procedure:

Identify the following symbols:

(symbols to identify with blank lines for answers)

Observations:

1. Why are map symbols important?

2. What are some other map symbols that you have seen?

38

CHAPTER 12
Careers in Soil Management

TEST YOUR KNOWLEDGE

Complete the following:

1. _____ _____ provide advice and assistance to farmers.
2. A soil conservationist may develop a farm _____ to allow the wise use of the land.
3. If erosion is a problem, the _____ is found and a plan to correct it is developed.
4. Soil conservationists might also:
   a.
   b.
   c.
   d.
5. _____ is the ability of the soil to allow water to move through it.
6. The soil conservation _____ is an assistant to a soil conservationist.
7. _____ _____ deal with the physical, chemical, and biological nature of the soil.
8. A _____ _____ is a collection of soil type maps and related information.
9. A _____ _____ analyzes soils for construction sites, lake sites, and other conservation activities.
10. A soil engineer must evaluate:
    a.
    b.
    c.
    d.
    e.
11. Percolation rate refers to the _____ at which water soaks into the ground.

CHAPTER 12
Careers in Soil Management

ACTIVITY 1

Purpose:

Determine length of pace.

Materials Needed:

100 foot tape

Research:

Pacing is one method of measuring distance. It consists of determining the average length of normal steps or paces. The length of a pace varies with the individual, the rate of speed, and the terrain.

Procedure:

1. Measure a distance of 100 feet.

2. Pace its length at least four times.

3. Divide the number of paces by four to determine the average number of paces in 100 feet.

Practice pace	Number of paces
1	
2	
3	
4	
Total	

Observations:

How many feet is your pace?

$$\frac{\text{Total number of paces } \_\_\_\_}{4} = \text{(average number of paces)}$$

$$\frac{100 \text{ feet}}{\text{(average number of paces)}} = \_\_\_\_ \text{ feet/pace}$$

CHAPTER 12
Careers in Soil Management

ACTIVITY 2

Purpose:

Describe the purpose of Natural Resources Conservation Districts.

Research:

Locate the following information, provided by the instructor, about

conservation districts: *their origin, development, and functions.*

Procedure:

Write a paragraph describing the role of the Natural Resources Conservation Districts.

Observations:

1. How might the Natural Resources Conservation District in your area assist you?

2. How might you become involved with your local Natural Resources Conservation District?

Job Exercise for CHAPTERS 4–12

Purpose:
Classify groups of rocks.

Materials Needed:
Rock and Mineral Kit
Rock and Mineral Study Guide

Research:
By using the materials, learn to identify the rocks listed here:

- Conglomerate
- Serpentinite
- Marble
- Limestone
- Pumice
- Obsidian
- Slate
- Anthracite Coal
- Calcareous Tufa
- Shale
- Basalt
- Talc Schist
- Scoria
- Sandstone
- Granite

Procedure:
1. Define igneous rock.

2. Define sedimentary rock.

3. Define metamorphic rock.

Observations:
1. Determine what group each rock belongs to.

2. Complete the table by placing a check mark in the appropriate box.

Rock	Igneous	Sedimentary	Metamorphic
Conglomerate			
Serpentinite			
Marble			
Limestone			
Pumice			
Obsidian			
Slate			
Anthracite Coal			
Calcareous Tufa			
Shale			
Basalt			
Talc Schist			
Scoria			
Sandstone			
Granite			

SECTION III
Water Resources

CHAPTER 13
Water Supply and Water Users

TEST YOUR KNOWLEDGE

Complete the following:

1. Water covers about _____ percent of the earth's surface.
2. _____ percent of the earth's water is located in the oceans, _____ percent is freshwater, and _____ percent is frozen in glaciers.
3. The amount of water on this planet is fairly _____.
4. The moving of water from place to place is called the water cycle or the _____ cycle.
5. The movement of water into the atmosphere is called _____.
6. Water that returns to the ocean is called _____ water.
7. Meteoric water can be _____, _____, or sleet.
8. _____ is the movement of water through the roots of plants and out the stomata.
9. In animals, water:
    a.
    b.
    c.
10. Lakes, ponds, and streams make up the _____ water.
11. Movement of surface water is called _____ water.
12. Water that soaks into the soil is called _____.
13. The main components of the hydrologic cycle include:
    a.
    b.
    c.
    d.
    e.
14. The main users of water include:
    a.
    b.
    c.
    d.
    e.
    f.
15. The main agricultural use of water is _____.
16. The most common methods of irrigation are _____ and _____.

17. Irrigation water not being drawn from groundwater must come from _____ water.
18. When irrigation water evaporates, _____ and _____ are left behind.
19. Industry uses more _____ than any other raw product.
20. Industry draws about _____ _____ gallons of water per day.
21. Each person uses about _____ gallons of water per day.

CHAPTER 13
Water Supply and Water Users

ACTIVITY

Purpose:

Identify surface water supplies.

Research:

1. Locate a state map.

2. Locate the following surface water areas in your area detailed on the map:

    a. Lakes
    b. Ponds
    c. Rivers
    d. Reservoirs
    e. Streams

Procedure:

Complete the following table by listing names of the surface water in your area.

Lakes	Rivers	Streams	Ponds	Reservoirs

Observations:

1. What is the main source of surface water supplies in your area?

2. Considering the main source of surface water, what is the name of the major supplier of this surface water?

CHAPTER 14
Water Pollution

TEST YOUR KNOWLEDGE

Complete the following:

1. For years, water was used as a means of disposing of _____.
2. Three main sources of water pollution include:
   a.
   b.
   c.
3. Pollution from cities affects not only the surface water runoff but also the _____ supply.
4. _____ and _____ are potential sources of groundwater pollution.
5. Industrial pollution includes:
   a.
   b.
   c.
   d.
6. _____ pollution involves returning heated water to a stream or river.
7. Warm water affects fish:
   a.
   b.
   c.
8. Techniques used to control thermal pollution are:
   a.
   b.
9. _____ _____ emit radiation as a result of the disintegration of their atomic nuclei.
10. Fish and wildlife can accumulate radioactive materials in their _____.
11. Organic wastes cause bacteria to place a demand on the _____ in the water.
12. The _____ test is used to determine the oxygen demand on water.
13. The most common agricultural pollutants include:
    a.
    b.
    c.
    d.
14. Nutrients that reach ponds make the water excessively rich in a process called _____.
15. Erosion causes topsoil to move into _____ and _____.

CHAPTER 14
Water Pollution

ACTIVITY

Purpose:

Determine pollution sources.

Research:

Review the three major water pollution groups. Water pollution can be separated by its source, into two categories—point-source and nonpoint-source pollution. Point-source pollution includes materials that are discharged directly from a specific source. Nonpoint-source includes runoff from sources that are widespread and harder to identify.

Procedure:

Complete the following table using the following information:

Source:	**Pollution Group:**
PS — Point source	U — Urban
NP — Nonpoint source (Diffuse)	I — Industrial
	A — Agricultural

Pollution	Source	Group
Detergents used to clean roads		
Sediments from fields		
Radioactive waste from a medical lab		
Salts to melt snow on roads		
Heated water from a power plant		
Feedlot waste		
Fertilizer applied to fields		
Pesticide residue from fields		
Radioactive waste from a nuclear power plant		
Landfill		
Waste from production of detergent		
Municipal dumps		
Waste from production of fertilizer		
Municipal sewage-treatment discharge		
Failing septic tanks		
Abandoned mine drainage		
Ocean dumping of city wastes		

Observations:
1. Of the types of pollution listed, which source occurred most often?
2. Of the types of pollution listed, which group occurred most often?

CHAPTER 15
Water Purification and Wastewater Treatment

TEST YOUR KNOWLEDGE

Complete the following:

1. There are about _____ cubic miles of water.
2. We do not "use water up," we remove it from the water cycle and then _____ it to the water cycle.
3. Once the water has been used, it must be _____ before it can safely return to the hydrologic cycle.
4. List the two basic wastewater treatment systems:
   a.
   b.
5. In nature water is never _____ and clean.
6. _____ are substances that dissolve other substances.
7. When one substance is dissolved in another, it is a _____.
8. Before we use water for human consumption, _____ must be removed.
9. List the three categories of impurities:
   a.
   b.
   c.
10. Chemical impurities can result in these three conditions:
    a.
    b.
    c.
11. The pH of water should be near _____.
12. Water with a pH higher than 7.0 is _____.
13. Very alkaline water is _____.
14. When water with _____ and manganese is exposed to air, it causes red or brown stains.
15. _____ water is a result of excess calcium or magnesium.
16. When sulfur is exposed to oxygen in water, it forms _____ _____ gas.
17. Excessive nitrogen and phosphorus promote the growth of _____ in water.
18. _____ materials in water must be removed or neutralized.
19. _____ are tiny green plants.
20. Bacteria and fungi can serve to break down _____ matter into harmless compounds.
21. _____ are single-celled animals that occur in water.
22. _____ refers to solid matter suspended in a liquid.

23. The objective of water treatment is to produce a _____ water supply.
24. Most _____ _____ is safe to drink as it comes from the ground.
25. Water that moves into a treatment system is referred to as _____.
26. Water that comes out of a treatment system is called _____.
27. A typical septic system consists of these three parts:
    a.
    b.
    c.
28. Wastewater treatment is divided into these three phases:
    a.
    b.
    c.
29. The primary system removes about _____ of the wastes from water.
30. The secondary waste treatment system involves biological processing of _____.
31. The tertiary waste treatment is the _____ processing of sewage wastewater.

CHAPTER 15
Water Purification and Wastewater Treatment

ACTIVITY

Purpose:
Interpret waste analysis results.

Research:
Review the standards detailed here:

Standards	
Total Suspended Solids	less than 45 mg/l
Biological Oxygen Demand	less than 45 mg/l
Ammonia	less than 12.75 mg/l
pH	less than 6.5 to 8.3
Dissolved Oxygen	more than 6.0
Total Residual Chlorine	less than .13 mg/l

Water released from the wastewater treatment plant must not exceed these standards.

Procedure:
Complete the table by indicating which of the following samples are not safe for release into the stream.

Information	Sample #1	Sample #2	Sample #3
Total Suspended Solids	50 mg/l	16 mg/l	25 mg/l
Biological Oxygen Demand	50 mg/l	13 mg/l	17 mg/l
Ammonia	2.67 mg/l	1.57 mg/l	15 mg/l
pH	7.3	7.34	8.1
Dissolved Oxygen	6.0 mg/l	7.38 mg/l	6.0 mg/l
Total Residual Chlorine	.07 mg/l	0.02 mg/l	.07 mg/l

Observations:

List the sample number that does not meet the standards and indicate which standards are not met.

Sample	Standards

CHAPTER 16
Water-Use Planning

TEST YOUR KNOWLEDGE

Complete the following:

1. Dams, reservoirs, and ponds have two basic functions:
   a.
   b.
2. Water management also involves the use of water to move _____.
3. The most common water transported items are _____ and _____.
4. The largest bodies of water are the _____.
5. Because of its high _____ content, ocean water in not useful in its present state.
6. The process of salt extraction is called _____.
7. Two methods of salt removal are:
   a.
   b.
8. The methods of membrane desalination are:
   a.
   b.
9. Reverse osmosis applies _____ to the salt water to make it flow from a salt solution to a fresh solution.
10. An electric current can cause calcium and chlorine to be separated in the _____ process.
11. The _____ process works because heated water evaporates and leaves behind any solids.
12. The three main distillation processes are:
    a.
    b.
    c.
13. The main problems involved with recycling water include:
    a.
    b.
    c.
    d.
14. The most common method of making rain is to seed clouds with _____ _____.

CHAPTER 16
Water-Use Planning

ACTIVITY

Purpose:

Calculate residential water rates.

Research:

Residential and commercial water rates differ because of the amount of water used, as shown:

0–300 cubic feet = $1.00/100 cubic feet
300+ cubic feet = $ .75/100 cubic feet.

Water rates decline as the total amount of water increases. Residential water bills may include garbage and sewer costs.

Review this example:

Consider a present reading of 15000 and a previous reading of 10000:

Present reading	150 (100 cu. ft.)
Previous reading	− 100 (100 cu. ft.)
	50 (100 cu. ft.)

$3 \times 1.00 = \$ 3.00$
$50 - 3 = 47 \times .75 = \$35.25$
$\overline{ \$38.25}$

Procedure:

Complete the following billing statements using the prices detailed in the Research section.

Calculate: 100 cu. ft. used, amount billed, total due.

---

① **MUNICIPAL UTILITY BILL**
CITY FINANCE OFFICE — UTILITY BILL DEPT.
222 MAIN ST. — SOMEWHERE, U.S.A. 12345
321-1234

SERVICE AT	714 WRIGHT CRT			
METER READING		100 CU. FT. USED	CODE	AMOUNT BILLED
PRESENT	PREVIOUS			
149	117		WC	
METER READING DATES			GC	4.00
073100	070100		SC	7.70

BILLS PAID AFTER DUE DATE SHOWN BELOW ARE SUBJECT TO A LATE PAYMENT CHARGE

DESCRIPTION OF CODES ON REVERSE SIDE

BALANCE FORWARD	AMOUNT BILLED	TOTAL DUE
ACCOUNT NUMBER	BILLING DATE	DUE DATE
033199006	08/28/00	09/22/00
LAST PMT. DATE	TOTAL PMTS. REC'D	LAST ADJ.
08/17/00	35.90	

KEEP THIS STUB FOR YOUR RECORDS

---

② **MUNICIPAL UTILITY BILL**
CITY FINANCE OFFICE — UTILITY BILL DEPT.
222 MAIN ST. — SOMEWHERE, U.S.A. 12345
321-1234

SERVICE AT	714 WRIGHT CRT			
METER READING		100 CU. FT. USED	CODE	AMOUNT BILLED
PRESENT	PREVIOUS			
149	100		WC	
METER READING DATES			GC	4.00
073100	070100		SC	7.70

BILLS PAID AFTER DUE DATE SHOWN BELOW ARE SUBJECT TO A LATE PAYMENT CHARGE

DESCRIPTION OF CODES ON REVERSE SIDE

BALANCE FORWARD	AMOUNT BILLED	TOTAL DUE
ACCOUNT NUMBER	BILLING DATE	DUE DATE
033199006	08/28/00	09/22/00
LAST PMT. DATE	TOTAL PMTS. REC'D	LAST ADJ.
08/17/00	35.90	

KEEP THIS STUB FOR YOUR RECORDS

---

③ **MUNICIPAL UTILITY BILL**
CITY FINANCE OFFICE — UTILITY BILL DEPT.
222 MAIN ST. — SOMEWHERE, U.S.A. 12345
321-1234

SERVICE AT	714 WRIGHT CRT			
METER READING		100 CU. FT. USED	CODE	AMOUNT BILLED
PRESENT	PREVIOUS			
345	75		WC	
METER READING DATES			GC	4.00
073100	070100		SC	7.70

BILLS PAID AFTER DUE DATE SHOWN BELOW ARE SUBJECT TO A LATE PAYMENT CHARGE

DESCRIPTION OF CODES ON REVERSE SIDE

BALANCE FORWARD	AMOUNT BILLED	TOTAL DUE
ACCOUNT NUMBER	BILLING DATE	DUE DATE
033199006	08/28/00	09/22/00
LAST PMT. DATE	TOTAL PMTS. REC'D	LAST ADJ.
08/17/00	35.90	

KEEP THIS STUB FOR YOUR RECORDS

---

④ **MUNICIPAL UTILITY BILL**
CITY FINANCE OFFICE — UTILITY BILL DEPT.
222 MAIN ST. — SOMEWHERE, U.S.A. 12345
321-1234

SERVICE AT	714 WRIGHT CRT			
METER READING		100 CU. FT. USED	CODE	AMOUNT BILLED
PRESENT	PREVIOUS			
555	65		WC	
METER READING DATES			GC	4.00
073100	070100		SC	7.70

BILLS PAID AFTER DUE DATE SHOWN BELOW ARE SUBJECT TO A LATE PAYMENT CHARGE

DESCRIPTION OF CODES ON REVERSE SIDE

BALANCE FORWARD	AMOUNT BILLED	TOTAL DUE
ACCOUNT NUMBER	BILLING DATE	DUE DATE
033199006	08/28/00	09/22/00
LAST PMT. DATE	TOTAL PMTS. REC'D	LAST ADJ.
08/17/00	35.90	

KEEP THIS STUB FOR YOUR RECORDS

CHAPTER 17
Careers in Water Management

TEST YOUR KNOWLEDGE

Complete the following:

1. The four main environmental science occupations are _____, _____, _____, and _____.
2. _____ study the structure, composition, and history of the earth's surface.
3. A groundwater geologist studies the _____ and _____ of underground water resources.
4. Many times geologists work _____.
5. Demand for geologists is expected to _____.
6. Aquatic science is the study of _____ waters.
7. _____ do research in developing fisheries and mining the ocean.
8. _____ involves studying the air that surrounds the earth.
9. As pressure increases to treat wastewater, the job of the _____ _____ plant operator becomes more important.
10. Employment in wastewater treatment is expected to _____.

CHAPTER 17
Careers in Water Management

ACTIVITY

Purpose:

Determine amount of water used.

Procedure:

Complete the following:

1. The average person will take a shower two times per day.

2. Two showers per day × 30 gallons = _____ gallons per day.

3. _____ gallons/day × 30 days per month = _____ gallons per month.

4. _____ gallons/month × 12 months per year = _____ gallons per year.

5. _____ gallons/year ÷ 7.46 gallons per cubic foot = _____ cubic feet.

Observations:

1. How many gallons per year are used by the average person taking a shower?

2. How many cubic feet per year are used by the average person taking a shower?

Job Exercise for CHAPTERS 13–17

Purpose:
Set up a sprinkler system.

Materials Needed:
valve
elbow
adaptor
¾-inch PVC pipe
tee
riser
head
PVC cement
PVC pipe cutter

Procedure:

1. Cut two 1-foot sections and a 2-foot section of *¾-inch PVC pipe*, using the *PVC pipe cutter*.

2. Screw the *adaptor* into the valve outlet.

3. Install *adaptor* to the end of a 1-foot section of *¾-inch PVC pipe* and an *elbow* at the other end.

4. Connect the 2-foot section of *¾-inch PVC pipe* to the *elbow*.

5. Connect the *tee* to the other end of the 2-foot section of pipe.

6. Connect the *riser* to the leg of the *tee*.

7. Screw the *head* to the *riser*.

8. Connect the other 1-foot section to the outlet of the *tee*, and connect the cap to the end of the *¾-inch PVC pipe*.

9. Once the system is complete, check with the instructor before gluing with *PVC cement*.

Observations:

Compare your completed system to the following picture:

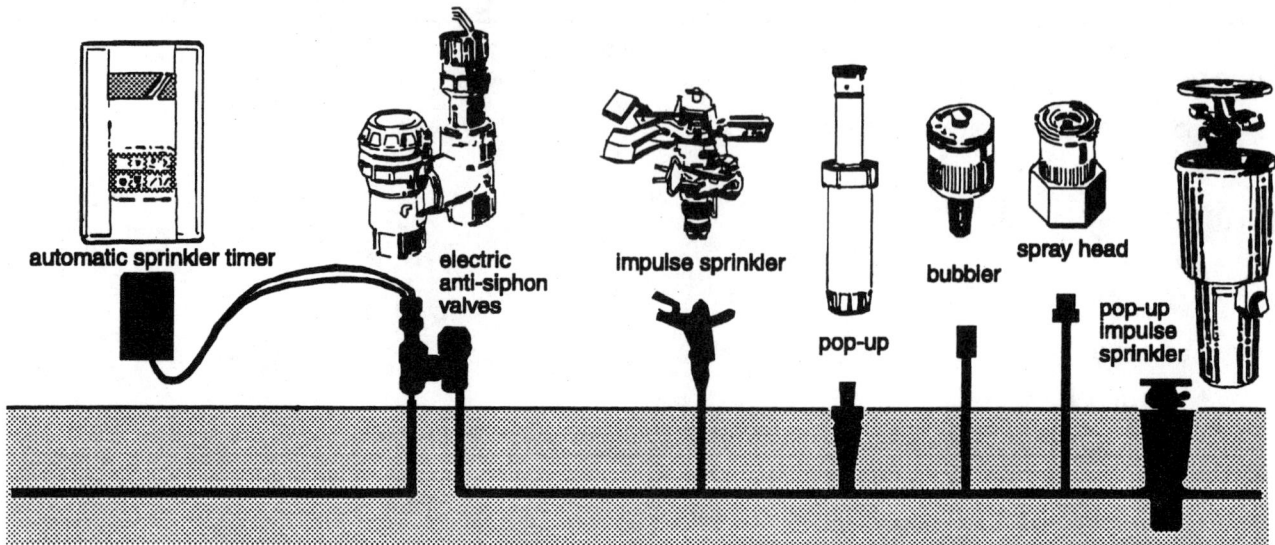

SECTION IV
Forest Resources

CHAPTER 18
Our Forests and Their Products

TEST YOUR KNOWLEDGE

Complete the following:

1. Forests are one of our _____ resources.
2. A _____ is a very complex community of trees, shrubs, plants, and animals.
3. Trees were the _____ of early settlers to America.
4. The _____ building industry of Europe depended upon the colonies for straight trees for their ships.
5. America has _____ billion acres of forests.
6. Only _____ million acres of America's forest are considered commercial.
7. _____ forests mean that the land is capable of producing economically useful forest.
8. Your state has _____ million acres of forests.
9. The American Forest Foundation System can give you information about establishing a _____ tree farm.
10. _____, _____ type, and _____ affect the kinds of tree that will grow in the United States.
11. The six major forest regions are:
    a.
    b.
    c.
    d.
    e.
    f.
12. The three main parts of a tree are:
    a.
    b.
    c.
13. The root system serves to _____ and _____ the tree.
14. Other functions of the root system include:
    a.
    b.
    c.
    d.
    e.

15. Types of roots include:
    a.
    b.
    c.
    d.
16. Roots grow both in _____ and _____.
17. Root growth happens by _____ division.
18. Root _____ absorb the water and nutrients taken in by the root system.
19. The trunk _____ the crown of the tree.
20. The trunk transports the _____ and _____ _____ upward from the root system.
21. A tree trunk consists of five parts:
    a.
    b.
    c.
    d.
    e.
22. The _____ consists of woody cells that are dead, which adds strength and stiffness to the tree.
23. The _____ is a thin layer of active cells that divides to produce new cells.
24. Outside the _____ layer is the phloem, which moves the food in the leaves downward to the roots and to the rest of the tree.
25. In the _____, new sapwood cells are large and soft walled.
26. In the _____, new cells are smaller and darker in color.
27. During a year with an extremely good growing season, the annual _____ will be very wide.
28. The crown of the tree includes _____, _____, _____, and _____.
29. Leaves convert water and carbon dioxide in the presence of sunlight into sugar in a process called _____.
30. _____ is the evaporation of water from the tree leaves.
31. _____ trees grow satisfactorily without direct sunlight.
32. Shade-intolerant trees need _____ sunlight.
33. _____ forests have trees of a single age and size.
34. _____ forests include trees in two or more size groupings.
35. The _____ of a forest is made up of the crowns of the taller trees.
36. Members of the forest canopy include:
    a.
    b.
    c.
    d.

37. Lumber grades include:
    a.
    b.
    c.
    d.
    e.
38. _____ woods include products such as paper, pulp, wood fiber, charcoal, explosives, and plastics.
39. Other forest-produced benefits include:
    a.
    b.
    c.
40. Forests contribute to the balancing of _____ and _____ _____ in the atmosphere.
41. Water and soil conservation of forests:
    a.
    b.
    c.

CHAPTER 18
Our Forests and Their Products

ACTIVITY

Purpose:

To identify the parts of a tree.

Procedure:

1. Review "Trees and Their Growth" in the text.

2. Complete the following pictures.

Observations:

1. Study the parts of the tree.

2. Study the parts of tree branches and roots.

3. Study the parts of the tree trunk.

Complete the following:

A.

B.

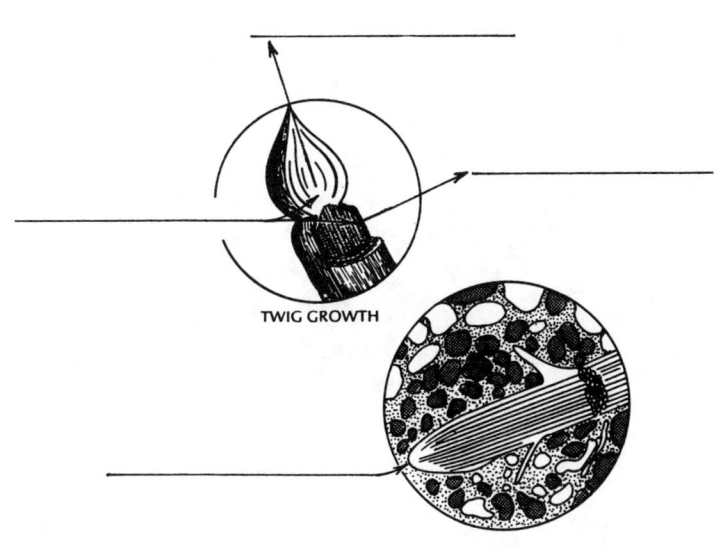

C.

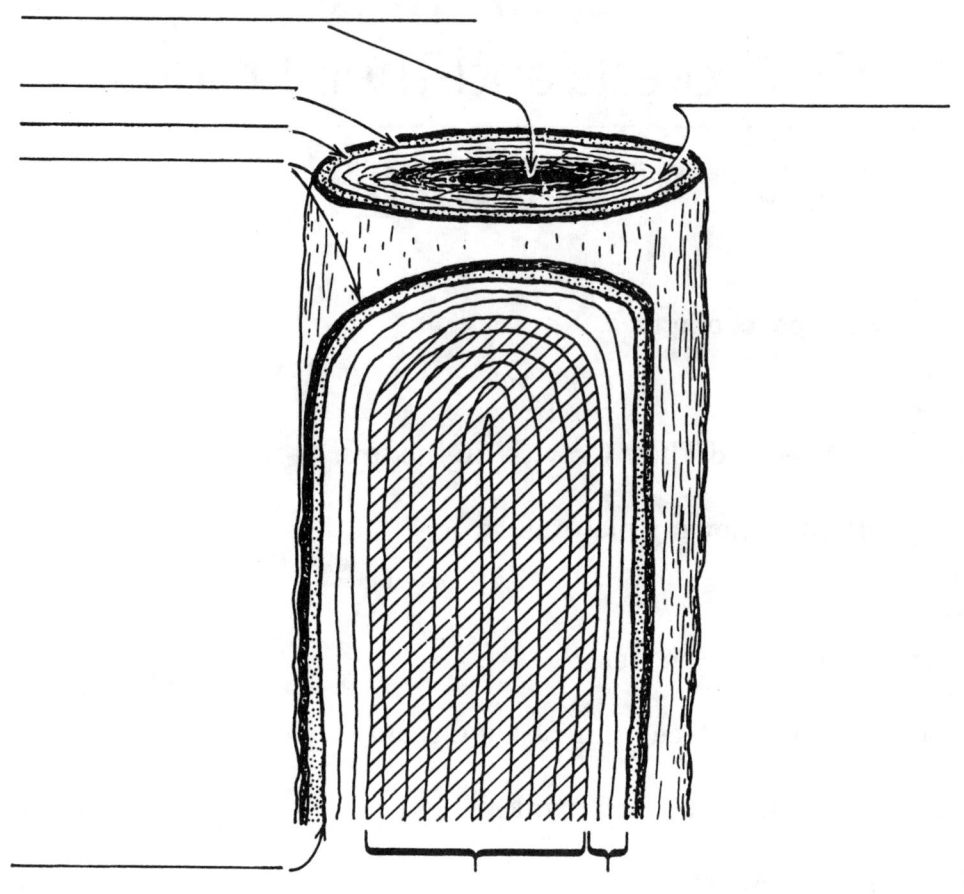

CHAPTER 19
Woodland Management

TEST YOUR KNOWLEDGE

Complete the following:

1. The standard unit of measure for most lumber is the _____ foot.
2. Timber and lumber are usually sold at a _____ per board foot.
3. A board foot is a piece of _____ wood one foot long, one foot wide, and one inch thick.
4. The board shape may change but the board foot volume must always be _____ cubic inches.
5. The formula for _____ feet is length times width times thickness, divided by 144.
6. A _____ foot is the amount of wood that fills a space one foot wide, one foot thick, and one foot high.
7. A standard _____ of wood is a stack, four feet by four feet by eight feet.
8. A cord of wood is _____ cubic feet.
9. Gross weight is used to measure _____.
10. The _____ of wood in a log is determined by its diameter and length.
11. DBH is the diameter of a tree at _____ feet above the ground level.
12. DBH is usually rounded to the nearest _____ class.
13. The three most common dendrometers are:
    a.
    b.
    c.
14. Tree height includes only the _____ length of the trunk.
15. When measuring saw timber, tree height is measured in _____ logs.
16. Pulpwood tree height is measured in _____.
17. Tree height is measured from the height of the _____ to the point where the cutoff diameter is estimated.
18. Types of hypsometers are:
    a.
    b.
    c.
    d.
19. Log _____ or tables are used to determine timber volume of a tree.
20. Estimating _____ timber volume is known as cruising timber.
21. Cuttings taken during the time from _____ to _____ are called intermediate cuttings.
22. Intermediate cuttings to assist young seedlings or saplings are called _____.
23. _____ cuttings in older trees are called improvement cuttings.

24. Removing trees from a stand with too many and too close trees is called _____.
25. Removing dominant trees is called _____.
26. _____ cutting removes injured, diseased, or insect-infested trees.
27. Removing damaged trees is called _____ cutting.
28. Controlled _____ is the clearing of undergrowth in a pine forest by fire.
29. In _____ cutting, trees are selected based on maturity, size, species, growth rates, and other factors.
30. When a forest is harvested in two or three stages, it is called _____ cutting.
31. _____ cutting involves removing the entire stand while leaving only selected seed trees.
32. In _____, all the trees of a tract are harvested in a single operation.
33. The four methods of reproducing the forest include:
    a.
    b.
    c.
    d.
34. A good forest management plan:
    a.
    b.
    c.
    d.

CHAPTER 19
Woodland Management

ACTIVITY

Purpose:

To determine the volume of a log.

Research:

1. Volume tables are used to determine the volume of a standing tree or cut log.

2. Volume tables can show cubic feet or board feet for an area.

3. Tables have the height across the top, and DBH (diameter at breast height) on the left side.

Procedure:

Review the table on the next page and practice using it.

Observations:

Complete the following using a board feet volume table.

Tree	Height in feet	DBH in inches	Board feet
1	100	24	
2	80	16	
3	75	15	
4	95	20	
5	90	25	

Gross Volume per Tree for Ponderosa Pine in the Black Hills
Board Feet in Scribner Rule

Board Feet Inside Bark
Merchantable Stem Excluding Stump and Top

Top Diameter — 6.0 Inches
Stump Height — 1.0 Foot

TOTAL HEIGHT

DBH	20	25	30	35	40	45	50	55	60	65	70	75	80	85	90	95	100	105	110	115	120	125
8	1	2	4	6	9	11	13	15	17	19	22	24	26									
9	3	6	8	11	14	16	19	22	24	27	30	33	35	38								
10	6	9	13	20	27	33	40	47	54	61	67	74	81	88	95							
11	8	12	21	29	37	45	53	61	69	78	86	94	102	110	118	127						
12	9	19	29	38	48	58	67	77	87	96	106	115	125	135	144	154	164					
13	15	27	38	49	60	72	83	94	105	117	128	139	150	161	173	188	203	218				
14	22	35	48	61	74	87	100	112	125	138	151	164	179	196	214	231	248	265	282			
15	29	44	58	73	88	103	118	133	147	162	179	199	218	238	257	277	297	316	336	355		
16	36	53	70	87	104	120	137	154	171	193	215	237	260	282	304	326	348	371	393	415	437	
17	44	63	82	101	120	139	158	179	204	229	254	279	304	329	354	379	404	429	454	479	504	529
18	53	74	95	116	138	159	183	211	239	267	295	323	350	378	406	434	462	490	518	546	574	602
19	62	86	109	133	156	183	214	245	276	307	338	369	400	431	462	493	524	555	586	617	648	679
20	72	98	124	150	178	212	246	281	315	349	383	418	452	486	521	555	589	623	658	692	726	761
21	82	111	139	167	205	243	280	318	356	394	431	469	507	544	582	620	658	695	733	771	808	846
22		124	155	192	234	275	316	357	399	440	481	523	564	605	647	688	729	770	812	853	894	936
23			173	218	263	309	354	399	444	489	534	579	624	669	714	759	804	849	894	939	984	1029
24				246	295	344	393	442	491	540	589	638	686	735	784	833	882	931	980	1029	1078	1127
25					327	380	433	486	539	593	646	699	752	805	858	911	964	1017	1070	1123	1176	1229
26						418	476	533	590	648	705	762	819	877	934	991	1049	1106	1163	1221	1278	1335
27							520	581	643	705	766	828	890	952	1013	1075	1137	1198	1260	1322	1384	1445
28								632	698	764	830	897	963	1029	1095	1162	1228	1294	1361	1427	1493	1559
29									755	826	897	968	1039	1110	1181	1252	1323	1394	1465	1536	1607	1678
30										889	965	1041	1117	1193	1269	1345	1420	1496	1572	1648	1724	1800

CHAPTER 20
Forest Enemies and Their Control

TEST YOUR KNOWLEDGE

Complete the following:

1. _____ kill more trees than any other enemy.
2. The main insect types that cause damage include:
   a.
   b.
   c.
   d.
   e.
   f.
   g.
   h.
3. Controls of insects include:
   a.
   b.
   c.
   d.
4. As humans change the forest, _____ controls may no longer work well.
5. Using a pathogen that causes a disease in insects is an example of _____ control.
6. Harvesting _____ trees is an example of management control.
7. Forest _____ is the study of tree diseases.
8. Noninfectious diseases are usually caused by _____ problems.
9. Infectious diseases are usually caused by _____.
10. Fungus diseases attack the leaves, stems, or _____ of the tree.
11. Controls of forest disease involve good _____.
12. Fomes rot is controlled by dusting stumps with borax or _____.
13. Trees damaged by stem fungus should be pruned or _____.
14. Crowded trees are more susceptible to disease and should be _____.
15. _____ resistant varieties should be planted.
16. _____ can be used to kill insects that spread disease.
17. When wildlife populations _____, damage to the forest may increase.
18. Overgrazing can cause _____ damage to the forest.

CHAPTER 20
Forest Enemies and Their Control

ACTIVITY

Purpose:

To identify forest insects and their damage.

Research:

1. Review the following list of insects:

    Aphid
    Scale
    Budworm
    Spider mite
    Beetle
    Borer
    Sawfly

2. Go to your local library and research information on the insects listed in item 1.

3. Review information about these insects.

Procedure:

Complete the following table:

Insect	Identification	Insect damage
Aphid		
Scale		
Budworm		
Spider mite		
Beetle		
Borer		
Sawfly		

Observations:

1. Which of the insects cause damage to pine trees?

2. Which of the insects cause damage that is not visible on the exterior of the tree?

CHAPTER 21
Fire!

TEST YOUR KNOWLEDGE

Complete the following:

1. Native Americans used fire as a way to clear the land and to improve _____.
2. Benefits of a prescribed fire include:
   a. reduces hazard of _____
   b. ready for _____ and _____
   c. improves _____ habitats
   d. removes undesirable _____ and _____
   e. controls forest _____
   f. improves quality of _____ for grazing
   g. improves _____ of the forest
   h. makes _____ to the forest easier
3. How does a prescribed fire work?
   a. _____ closely
   b. do not allow to get too _____
   c. do not allow to get out of _____
4. The largest fire in U.S. history was the _____, Wisconsin, fire.
5. In order for a fire triangle to occur, it must have _____, _____, and _____.
6. Most fires are caused by _____.
7. Causes of fire include:
   a.
   b.
   c.
   d.
   e.
   f.
   g.
8. The three categories of forest fires include:
   a.
   b.
   c.
9. An _____ is used to determine the direction from the tower to the fire.
10. With the help of a second tower, _____ is used to determine the fire location.

11. Factors that affect the anatomy of a fire include:
    a.
    b.
    c.
    d.
    e.
    f.
    g.
    h.
    i.
12. The National Fire Danger Rating Service predicts fire potential using:
    a.
    b.
    c.
13. The best method of fire suppression is _____ _____.
14. Removing fuel from the fire is a method of _____ attack on a fire.
15. A _____ _____ could be a lake, stream, or bulldozer-created fire line.
16. _____ are small fires set along the firebreak on the side toward the fire.
17. The operations involving patrolling of the fire line, watching for spot fires, and extinguishing small fires are called _____ up.

CHAPTER 21
Fire!

ACTIVITY

Purpose:

To identify forest tools and equipment.

Research:

1. Locate a Forestry Suppliers catalog.

2. Find each tool in the catalog index.

Procedure:

Complete the following table.

Tool/Equipment	Item number	Price
Tree stick		
Diameter tape		
Increment borer		
Bark gauge		
Tree calipers		
Forester axe		
Hand compass		
Plastic flagging		
Clinometer		
Safety hat		

Observations:

1. Which of the listed tools is the most expensive?

2. Which of the listed tools have you not used?

CHAPTER 22
Careers in Forestry

TEST YOUR KNOWLEDGE

Complete the following:

1. Most jobs in forestry are in the _____, _____, and marketing part of the industry.
2. A _____ is a professional who helps manage, direct, and protect our nation's forest resource.
3. Forester responsibilities include:
   a.
   b.
   c.
   d.
   e.
   f.
   g.
   h.
   i.
   j.
4. There are about _____ professional foresters.
5. About _____ percent of foresters work for the federal government and 30 percent for state and local governments.
6. Employment opportunities for foresters are expected to _____ slowly.
7. A professional forester requires a _____ degree in forestry.
8. A forestry technician works under the supervision of the _____.
9. Much work of the forestry technician is _____ _____.
10. Forestry _____ may have completed a one- or two-year post-secondary training program.
11. The first task of a logger is _____ the trees.
12. An _____ removes a wedge of wood from the side where the logger wants the tree to fall.
13. When a tree is cut into logs or bolts it is called _____.
14. _____ involves dragging logs to a loading area.
15. Jobs in logging are expected to remain _____ as mechanization in the logging industry continues.
16. Forestry subjects are taught at three levels:
   a.
   b.
   c.

CHAPTER 22
Careers in Forestry

ACTIVITY

Purpose:

Identify the responsibilities of a forester.

Research:

1. Choose a forestry career.

2. Using various career sources, research and locate information about this favorite career.

Procedure:

List the sources used and the information obtained from these sources.

Observations:

1. What is the nature of work for this career?

2. What is the employment outlook for this career?

3. What training is required for this career?

Job Exercise for CHAPTERS 18–22

Purpose:
Measure diameter of a tree, (FF P 86-87) for use in volume calculations.

Materials Needed:
Diameter tape
Carpenters tape measure
Safety equipment as needed (gloves, hard hats, eye and ear protection)

Procedure:
1. Assemble the needed tools and equipment and carry to the site where trees are to be measured.

2. With the carpenter tape, measure to a height of 4.5 feet above ground level, on the uphill side of tree.

3. Secure the end of the diameter tape to the bark at the 4.5 foot height level.

4. Extend the diameter tape around the tree until it meets the end.

5. Read the diameter directly from the tape; tree diameters are usually recorded in the nearest two inch class (8", 10", 12").

6. Return tools and equipment to the storage area, and put away properly.

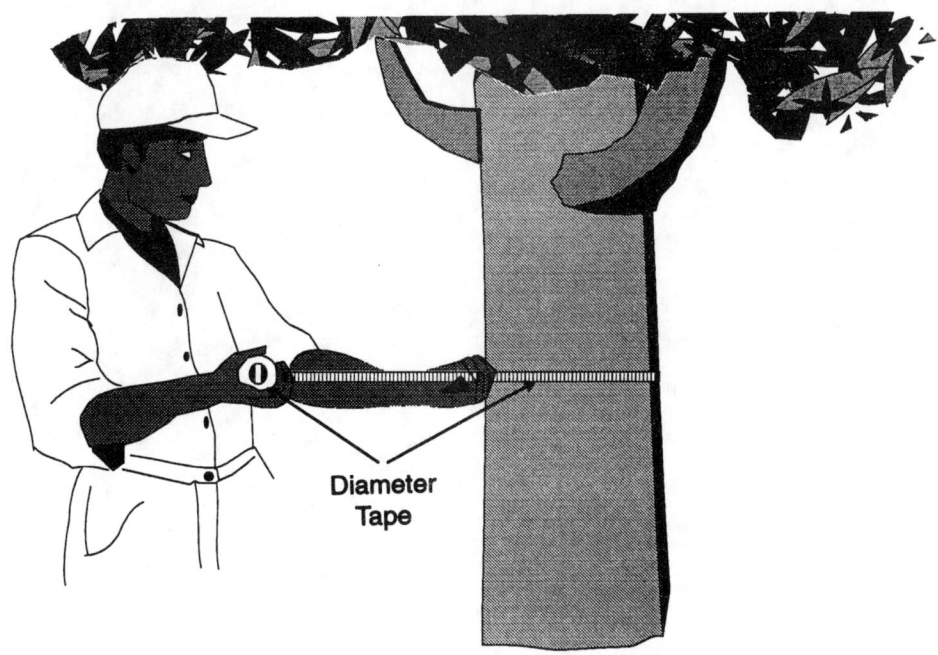

Observations:

Complete the following table for the trees you measured.

Tree	Diameter
1	
2	
3	
4	
5	

SECTION V
Fish and Wildlife Resources

CHAPTER 23
Fish and Wildlife in America

TEST YOUR KNOWLEDGE

Complete the following:

1. The definition for wildlife usually includes _____, _____, and _____.
2. America was established in the world _____ trade business at the expense of the wildlife.
3. Many wildlife species were killed because they appeared _____.
4. The Endangered Species Conservation Act called for the protection of fish and wildlife threatened with _____.
5. _____ species are species that no longer exist.
6. _____ species are no longer common and are in danger of becoming extinct.
7. The last known _____ pigeon died in 1914.
8. The _____ parakeet became extinct because their feathers were used for hats.
9. A _____ caused the death of the last protected heath hens.
10. Using their feathers for _____ caused the extinction of the Labrador duck.
11. There are 251 mammals on the U.S. Department of _____ endangered list.
12. Humans and _____ threaten the bighorn sheep population.
13. Fur collectors used _____ to hunt and kill polar bears.
14. The smallest white tail deer, the _____ deer, is protected from hunting.
15. Bounties on the _____ are allowed only if the population endangers the deer population.
16. Mountain lions are hunted by _____ in some parts of the United States.
17. There are _____ birds on the endangered species list.
18. The whooping cranes fly from Texas to Canada each year during _____.
19. Presently there are only about _____ whooping cranes.
20. Bounties were once placed on the bald eagles because they were thought to have fed on _____.
21. The ivory-billed woodpeckers were once thought to be _____.
22. The _____ _____ Foundation was formed to help manage the prairie chicken population.
23. There are _____ endangered fish species.

CHAPTER 23
Fish and Wildlife in America

ACTIVITY

Purpose:

Compile a profile of an endangered species.

Research:

Choose an endangered species you are interested in. Research this species by going to your local library and keeping notes.

Procedure:

1. Compile a one page report about your chosen rare or endangered species.

2. Include the following information:

    Description
    Habitat
    Reproduction
    Role of animal
    Future
    Summary

Observations:

1. What habitat is the most critical for this animal?

2. What would you do to improve the habitat for this animal?

CHAPTER 24
Game Management

TEST YOUR KNOWLEDGE

Complete the following:

1. The art of making land produce a sustained annual crop of wild game for recreational use is _____ management.
2. _____ includes the basic animal requirements of food, cover, territory, and home range.
3. _____ are plant eaters.
4. _____ are meat eaters.
5. _____ eat many different types of food.
6. _____ eat seeds.
7. _____ animals eat a great variety of food.
8. _____ animals eat a specialized diet.
9. _____ animals are more likely to starve in a food-scarce season.
10. _____ is a place that will protect game from harsh weather conditions.
11. Water is important in:
    a.
    b.
    c.
12. The area that the game travels is called its _____ _____.
13. The area an animal will defend is its _____.
14. Game management procedures include:
    a.
    b.
    c.
    d.
    e.
    f.
15. Land set aside for the protection of wildlife is called a _____, reserve, or wilderness area.
16. The most common habitat developments are _____ plantings, and woodland management.
17. Because of changing farming methods, _____ areas have become nonexistent.
18. Farmers are now asked to set aside a _____ area for wildlife feeding.
19. By not using woodland for grazing, farmers can see an _____ in game populations.
20. Rhode Island was the first state to establish a _____ hunting season.
21. The first bag limit was initiated in 1878 in _____.

86

22. Game populations depend on many factors:
    a.
    b.
    c.
    d.
    e.
23. One facet of game management has been to control _____.
24. Predators can be beneficial for the following reasons:
    a.
    b.
    c.
    d.
25. Bringing in new species is called the introduction of _____.
26. _____ _____ refers to the number of game animals in a defined area.
27. _____ _____ refers to the amount of game an area will provide the essentials of life for.
28. The management of wildlife belongs to the _____.
29. Federal legislation governing wildlife includes:
    a.
    b.
    c.
    d.
    e.
    f.

CHAPTER 24
Game Management

ACTIVITY

Purpose:

Determine carrying capacity.

Research:

Population density refers to the number of game animals in a defined area. Carrying capacity refers to the amount of game an area will provide the essentials of life for. A population is the number of some species living in a particular area. Carrying capacity of an area is influenced by many factors that cause a change in the habitat. As a result of these changes, the population levels rise and fall.

Procedure:

Graph the following comparison information of black-tailed deer.

Month	Good Range	Poor Range
DEC	62	26
MAY	50	25
JUNE	90	35
JULY	84	31
DEC	63	27

Observations:

Select true or false statements, by placing a T or F next to the statements.

_____ 1. The population is higher in the spring/summer.

_____ 2. The population is lower in the winter.

_____ 3. The average population level on good range is higher than on poor range.

_____ 4. Population levels continue to rise throughout the year.

CHAPTER 25
Marine Fisheries Management

TEST YOUR KNOWLEDGE

Complete the following:

1. The _____ can be considered the last frontier on this planet.
2. The four major physical parameters of the ocean are:
   a.
   b.
   c.
   d.
3. Ocean _____ is usually determined by depth and light penetration.
4. The five ocean depth zones are:
   a.
   b.
   c.
   d.
   e.
5. The supratidal and intertidal areas are _____ the water level.
6. The _____ zone is the part of the ocean where sunlight penetrates the water.
7. _____ refers to the concentration of salts within the ocean water.
8. The concentrations and types of _____ vary throughout the ocean.
9. _____ is defined as the number of grams of dissolved salt in 1,000 grams of sea water.
10. The average salinity of the ocean is _____ %.
11. _____ changes in the ocean occur as we move to different latitudes and different depths of the ocean.
12. The temperature stratification consists of three layers:
    a.
    b.
    c.
13. The higher the temperature, the _____ the density.
14. The density of the ocean _____ as the pressure and depth increase.
15. Water movements in the ocean include:
    a.
    b.
    c.

16. _____ generated waves are divided into sea, swell, and surf.
17. The _____ wave is found with temperature changes.
18. _____ waves are those with massive power caused by storms, hurricanes, and landslides.
19. The stationary wave moves only _____ and _____.
20. _____ are waves caused by the gravitational attraction of the sun and moon.
21. Currents are divided into surface current, turbidity currents, and _____ currents.
22. Marine animal life of the ocean can be divided into four groups:
    a.
    b.
    c.
    d.
23. The most common microscopic animals are the _____.
24. _____ is the staple food for many species of fish.
25. The four major marine fish for market are:
    a.
    b.
    c.
    d.
26. Other fish of less economic importance are:
    a.
    b.
    c.
    d.
27. Of the salmon:
    a. The seven main species are:
        (1)
        (2)
        (3)
        (4)
        (5)
        (6)
        (7)
    b. length 16 in. to _____ in.
    c. weight 5 lbs. to _____ lbs.
    d. salmon begin life in fresh water, but migrate to _____ to live and grow
    e. the Atlantic salmon is the only species that does not die after _____
    f. eggs hatch in _____ months
28. Management techniques include:
    a.
    b.
    c.
    d.

29. Salmon are sold:
    a.
    b.
    c.
    d.
30. Tuna:
    a. is a member of the _____ family
    b. weighs up to _____ lbs.
31. Tuna are caught by:
    a.
    b.
    c.
32. _____ _____ may cause the netting of porpoises, which is unlawful.
33. _____ fish are used in products such as livestock feed, soup, and fertilizer.
34. Shrimp is gathered by use of nets called _____.
35. The oysters of the Persian Gulf and the Pacific Ocean are responsible for making _____.
36. The largest crab is the _____ king crab.
37. Lobsters are caught in traps called _____.
38. The main mammals of the ocean include:
    a.
    b.
    c.
    d.
39. _____ whales obtain their food by straining plankton.
40. Toothed whales have a lower set of _____ teeth.
41. The Whaling Commission has stopped the hunting of the following whales:
    a.
    b.
    c.
    d.
    e.
42. Seals are divided into three groups:
    a.
    b.
    c.
43. _____ are the only tusked seals.
44. The area where a freshwater source opens into the ocean is called an _____.
45. _____ is the artificial propagation of marine animals.
46. The most commonly farmed ocean crop is the _____.

CHAPTER 25
Marine Fisheries Management

ACTIVITY

Purpose:

Calculate dressing percent of fish.

Research:

Catfish usually yield 55 to 60 percent of their live weight in dressed form. The head, vicera, and skin are equal to about 40 to 45 percent of the weight of a whole catfish. The ideal size of a catfish for the retail market is a live weight of 1 lb. to 1.25 lbs. Dressed fish will weigh from 8 to 10 ounces dressed weight.

Example: Live weight = 10 ounces
Dressing percent   60%

Live weight × Dressing percent = Dressed weight
10   ×   .60   = 6 ounces

Procedure:

Complete the following table concerning dressing percent.

Live Weight	Dressing percent	Dressed weight
16 oz.	60	
18 oz.	60	
20 oz.	60	
16.5 oz.	55	
24.5 oz.	55	

Observations:

1. What are some factors that might affect the dressing percent?

2. What are some ways that the dressing percent may be increased?

CHAPTER 26
Freshwater Fishery Management

TEST YOUR KNOWLEDGE

Complete the following:

1. Lakes can be divided into three zones:
   a.
   b.
   c.
2. The littoral zone is _____ and contains rooted vegetation.
3. Greenish-brown _____ inhabit the littoral zone.
4. The _____ zone contains no rooted vegetation, but phytoplankton is present.
5. Photosynthesis takes place in both the littoral zone and the _____ zone.
6. The bottom zone of the lake is the _____ zone.
7. _____ is the most common organism in the profundal zone.
8. The main uses of a farm pond include:
   a.
   b.
   c.
   d.
   e.
   f.
9. _____ ponds are impounded behind an earth embankment or dam.
10. _____ ponds are made by digging a pit below the surrounding ground level.
11. The _____ is land that drains into the pond.
12. The _____ assures that water will never flow over the dam.
13. The pond basin and pond banks should be planted with a _____ crop.
14. Trees and shrubs may be planted around the pond, but not too close to the _____.
15. The most common stocking includes:
    a.
    b.
    c.
16. Largemouth bass:
    a. have a large _____
    b. dark _____ or blotches on side
    c. spawn readily in _____ water
    d. eat aquatic _____
    e. take about _____ years to reach 12 inches long

f. record is _____
   g. nesting begins when water temperature reaches _____ degrees
17. Bluegill:
   a. a _____ source for bass
   b. eat _____
   c. take _____ growing seasons to mature
   d. spawn from May to _____
18. Channel catfish:
   a. feed on plants, insects, and small _____
   b. average life span is _____ _____
19. The main fish management procedures include:
   a.
   b.
   c.
   d.
   e.
   f.
20. _____ provides food, shelter, oxygen, and spawning and nesting habitats.
21. An overabundance of aquatic _____ can create problems.
22. The ways to control aquatic plant population include:
   a.
   b.
   c.
   d.
23. Fish sampling methods include:
   a.
   b.
   c.
   d.
24. Most ponds are equipped with a _____, which is necessary to completely empty the pond.
25. Fish may be removed by poisoning them with _____ or sodium sulfite.
26. Population adjustment procedures include:
   a.
   b.
   c.
27. Fertilizer can cause an increase production of _____, a principal food source for fish.
28. Nutrients include:
   a.
   b.
   c.
   d.
29. Increasing plant growth can cause winter fish kills, as bacteria that are decomposing waste use most of the _____ in the water.

30. The main factors associated with water quality include:
    a.
    b.
    c.
    d.
31. Fish grow best if the temperature is above _____ degrees.
32. Oxygen is added to water by:
    a.
    b.
    c.
33. The ideal pH for fish is between _____ and _____.
34. Fish are _____ if they grow in muddy water.

CHAPTER 26
Freshwater Fishery Management

ACTIVITY

Purpose:

Identify characteristics of fish.

Research:

The most common stocking fish include bass, bluegill, and catfish. Not all fish are suited for pond stocking. Before a decision is made concerning which species to stock, their biology should be understood.

Procedure:

Review the information about common freshwater fish in the text.

Observations:

Distinguish between characteristics of the common stocking fish by placing one of the following symbols next to the correct statement:

B — bass
Bl — bluegill
C — catfish

_____ 1. Have a large mouth
_____ 2. Feed on plants, insects, and small bluegill
_____ 3. Average life span is 6 to 8 years
_____ 4. Spawn from May to August
_____ 5. Record is 22 lbs. and 4 oz.
_____ 6. Dark stripe or blotches on side
_____ 7. Take about 3 years to reach 12 inches long
_____ 8. A food source for bass
_____ 9. Take three growing seasons to mature
_____ 10. Eat aquatic insects
_____ 11. Spawn readily in shallow water
_____ 12. Eat plankton

CHAPTER 27
Careers in Fish and Wildlife Management

TEST YOUR KNOWLEDGE

Complete the following:

1. The governmental watchdog of the game management program is the conservation officer or _____ _____.
2. The conservation officer's duties include:
   a.
   b.
   c.
   d.
   e.
   f.
   g.
   h.
   i.
3. The wildlife biologist deals with the _____ practices of game rather than the unlawful activities against game.
4. Wildlife biologists are responsible for:
   a.
   b.
   c.
   d.
   e.
   f.
   g.
   h.
   i.
   j.
5. A fish and wildlife technician:
   a.
   b.
   c.
6. An _____ _____ biologist works mainly with problems caused by rodents and predators.
7. A _____ biologist works with habitats, spawning, and artificially grown young.
8. Many conservation jobs require a college _____.

9. The employment opportunities offered by the state and federal governments are not expected to _____.
10. Starting salaries for game biologists are about _____ per year.
11. The main job of the fish _____ technician is to produce fish for food or game.
12. The three main areas of fish culture are:
    a.
    b.
    c.
13. A _____ may work with both domestic and wild animals.

CHAPTER 27
Careers in Fish and Wildlife Management

ACTIVITY

Purpose:
Take pH test

Materials Needed:
A pH test kit
Water samples

Research:
The ideal pH for fish is between 6.5 and 9.0. pH is the measure of relative amounts of acidity and alkalinity. A pH of 7.0 is neutral. A number less than 7.0 indicates that the water is acidic; a number greater than 7.0 indicates that the water is alkaline.

Procedure:

1. Locate a pH test kit.

2. Follow the instructions in the kit to make pH tests on the samples provided by the instructor.

3. Complete this table:

Sample	pH
1	
2	
3	
4	
5	

Observations:

1. Which of the samples would be acceptable for fish production?

2. How might you change a pH condition that is not acceptable?

Job Exercise for CHAPTERS 23–27

Purpose:
Complete an application for employment.

Procedure:
1. Always be prepared to fill out an application.
2. Look over the entire application before beginning to write.
3. Follow the directions on the form.
4. Write or print neatly.
5. Correct mistakes with a simple line through the word.
6. Be honest.
7. Answer all questions.
8. Make a copy of the application whenever possible.

Observations:
Complete the following application.

APPLICATION FOR EMPLOYMENT

Prospective employees will receive consideration without discrimination because of race, religion, color, sex, age, national origin or physical handicap.

Last Name	First	Middle	Date
Street Address			**Home Phone** ( )-
City, State, Zip			**Business Phone** ( )-

Are you over the age of 18? ☐ Yes ☐ No	If not, do you have valid working papers? ☐ Yes ☐ No	Position Desired:
Do you have any impairments — physical, mental or medical — which would interfere with your ability to perform the job for which you have applied?	☐ Yes If yes, please explain. ☐ No	Date Available for Work
Have you ever been convicted of a crime? ☐ Yes ☐ No	Charge, Date & Disposition:	Salary Expected
Are you legally eligible for employment in the U.S.? (Proof of citizenship or immigration status may be required upon employment.) ☐ Yes ☐ No		Have you ever been bonded? ☐ Yes ☐ No
How did you learn of our organization?	If the job requires, are you available to work overtime? ☐ Yes ☐ No	

Indicate skills you possess:

Typing Speed _____ wpm   Steno Speed _____ wpm

Computers (specify):   Software (specify):

List any professional memberships which you feel would be an asset for the position applied:

Military

Served in U.S. Armed Forces ☐ Yes ☐ No	Date Inducted:	Date Discharged:
Rank at Discharge	Briefly describe your duties:	

EDUCATION

	High School	College/University	Graduate/Professional	Other
School Name				
Years Completed: (circle)	9 10 11 12	1 2 3 4	1 2 3 4	1 2 3 4
Diploma/Degree				
Describe course of Study:				

EMPLOYMENT

1. Company Name _____ Telephone ( )–
   Address _____ ☐ Full Time ☐ Part Time
   Name of Supervisor/Manager _____ Type of Business:

	Starting	Final	Principal Duties:
Dates:			
Position:			
Salary:			
Reasons for leaving:			

2. Company Name _____ Telephone ( )–
   Address _____ ☐ Full Time ☐ Part Time
   Name of Supervisor/Manager _____ Type of Business:

	Starting	Final	Principal Duties:
Dates:			
Position:			
Salary:			
Reasons for leaving:			

3. Company Name _____ Telephone ( )–
   Address _____ ☐ Full Time ☐ Part Time
   Name of Supervisor/Manager _____ Type of Business:

	Starting	Final	Principal Duties:
Dates:			
Position:			
Salary:			
Reasons for leaving:			

4. Company Name _____ Telephone ( )–
   Address _____ ☐ Full Time ☐ Part Time
   Name of Supervisor/Manager _____ Type of Business:

	Starting	Final	Principal Duties:
Dates:			
Position:			
Salary:			
Reasons for leaving:			

We may contact the employers listed above unless you indicate otherwise.
DO NOT CONTACT: Employer Number(s): _____

FOR EMPLOYERS USE ONLY

REFERENCE CHECK

References Supplied: ☐ Yes ☐ No

Person/Organization Contacted	Results

Tests Administered	Score/Rating	Comments

Interviewer Name	Comments

Explain why you desire this position and why you believe you would do well at it.

I hereby declare that the information provided on this application, or in any attachments provided by me, is true and complete. I understand that if employed, any falsified information or omission of fact on this application may disqualify me from further consideration for employment and may be considered justification for dismissal if discovered at a later date.

I authorize Delmar to investigate the accuracy of any information from any person or organization (unless otherwise noted) and hereby release Delmar and all persons from all claims and liability of any nature arising from such investigation.

_____   _____
Signature of Applicant   Date

SECTION VI
Outdoor Recreation Resources

CHAPTER 28
Recreation on Public Lands

TEST YOUR KNOWLEDGE

Complete the following:

1. City dwellers visit outdoor recreation areas more so than do _____ people.
2. The federal government's main conservation programs are:
   a.
   b.
   c.
   d.
   e.
   f.
3. The National Park Service was established in _____.
4. The national parks are divided into three basic categories:
   a.
   b.
   c.
5. The park's system contains _____ areas designated as natural areas.
6. The national forests are _____ regulated than the national park's system.
7. There are _____ islands of 10 acres or more near the United States.
8. Over _____ million acres have been designated as wilderness areas.
9. Wilderness areas may be part of a national forest area if the area is roadless and totals more than _____ acres.
10. The purpose of the national _____ system is to provide outdoor recreation by the use of trails.
11. There are about _____ trails in the national trail system.
12. The national system of wild and scenic rivers classifies the rivers as:
    a.
    b.
    c.
13. State recreation areas include:
    a.
    b.
    c.
    d.
    e.
    f.

g.
   h.
   i.
   j.
14. It costs the government _____ of dollars to repair and clean areas vandalized by inconsiderate people.

CHAPTER 28
Recreation on Public Lands

ACTIVITY

Purpose:

Evaluate state recreation areas.

Research:

1. Locate state tourism information.

2. Find the areas detailing *state* recreation agencies.

Procedure:

Determine the name for the following state recreation agencies. Detail the information that can be found under the respective headings:

- State parks

- State recreation areas

- State forests

- State fish hatcheries

- State memorials

- State museums

- Nature preserves

- Historic preservation sites

- Fish and wildlife areas

Observations:

1. What agency would you contact for information for damage caused by deer?

2. What agency would you contact for information about camping in a state park?

3. What agency would you contact for information about federal endangered species?

110

CHAPTER 29
Outdoor Safety

TEST YOUR KNOWLEDGE

Complete the following:

1. Two-thirds of all gun casualties are caused by persons under the age of _____.
2. The ten commandments of gun safety include:
   a.
   b.
   c.
   d.
   e.
   f.
   g.
   h.
   i.
   j.
3. A _____ _____ means a hunter can take a certain number of animals per day or season.
4. Hunter _____ is a bright, fluorescent color that can be easily seen.
5. _____ and arrows are used for target practice and hunting.
6. The _____ is the end of the arrow that fits onto the bow string.
7. Much snowmobiling is done after _____.
8. All _____ should adhere to the basic snowmobile safety code.
9. A _____ should be aware of basic first-aid procedures.
10. _____ occurs when the body loses heat faster than it can be produced.
11. A _____ must contain lifesaving equipment.
12. The four kinds of personal flotation devices are:
    a.
    b.
    c.
    d.
13. The _____ is the front of the boat.
14. The common water sports include:
    a.
    b.
    c.
    d.
    e.

111

CHAPTER 29
Outdoor Safety

ACTIVITY

Purpose:

Determine your fire escape route.

Research:

Walk around your home and make notes on where the windows and doors (internal and external) are located on all levels.

Procedure:

1. Draw a floor plan of your household, including all levels, windows, and doors.

2. Establish one primary escape route.

3. Use arrows to indicate an escape route from each room in the house.

4. Example:

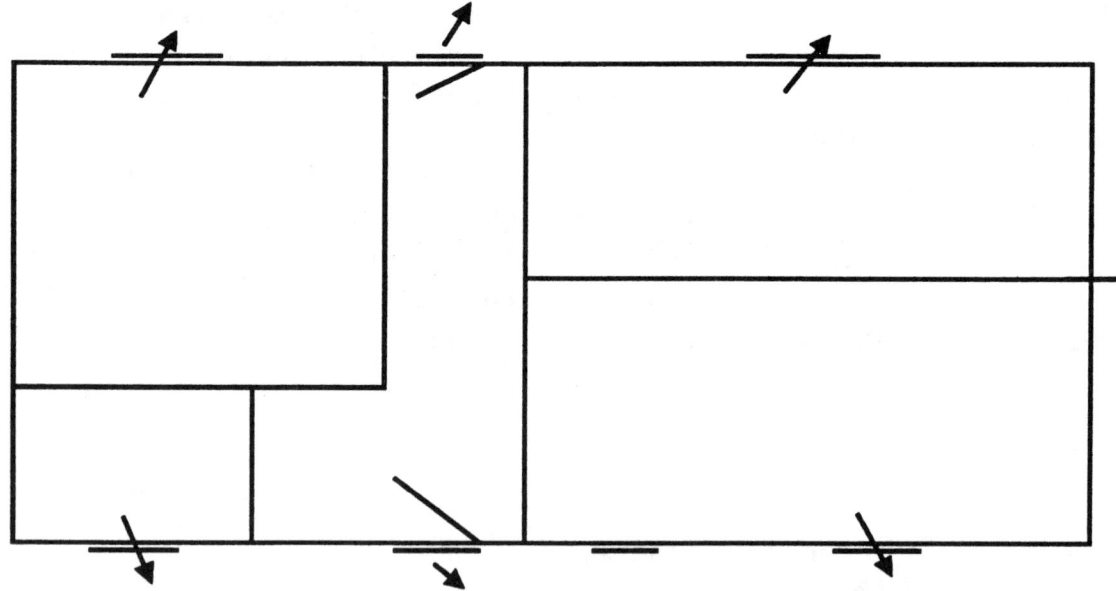

5. Make your plan drawing here:

Observations:

Mark an X in the location outside the house where everyone is to meet in case of a fire. Why is this an important step?

CHAPTER 30
Careers in Outdoor Recreation

TEST YOUR KNOWLEDGE

Complete the following:

1. Careers in outdoor recreation are _____ in number.
2. Outdoor recreation careers include _____, _____, skilled, or unskilled jobs.
3. Government employees in outdoor recreation include:
   a.
   b.
   c.
   d.
   e.
   f.
   g.
   h.
4. Park rangers:
   a. _____ the park in good condition
   b. conduct _____ programs
   c. guide _____
   d. help _____ the park
   e. _____ to local groups about park activities
   f. should like to work with _____
   g. must enjoy the _____
   h. may work in an _____
   i. must have at least a four-year _____ degree
5. Park aides and technicians:
   a. work on _____ programs
   b. enforce the _____
   c. operate _____
   d. fight _____
   e. work for the park _____
   f. require a high school _____
6. Park guides and guards:
   a. guides provide _____ to the public
   b. guards protect the _____

114

7. Fish and Wildlife guides:
   a. recreational programs have _____ in the past years
   b. most careers involve _____ visitors through their areas
8. Recreationists:
   a. plan, _____, and _____ recreational programs
   b. include a large number of _____, _____, and volunteer positions
9. Private recreation business opportunities include:
   a.
   b.
   c.
   d.
   e.
   f.
   g.
   h.
   i.
   j.
   k.
   l.
   m.
   n.

CHAPTER 30
Careers in Outdoor Recreation

ACTIVITY

Purpose:

Evaluate outdoor recreation regulations.

Research:

1. Locate the state game, fish, and parks hunting regulations and/or fishing regulations.

2. Locate the list of game animals, provided by the instructor.

Procedure:

1. Find the information about each animal in the handbook.

2. Determine the bag limit for each game animal.

3. Determine the season of each game animal.

4. Complete the following chart:

	Game Animal	Bag limit	Season
1			
2			
3			
4			
5			
6			
7			
8			
9			
10			
11			
12			

Observations:

1. Considering today's date, what is the next hunting season?

2. List four other items of hunting information found in the handbook:

    a.

    b.

    c.

    d.

3. List safety issues in hunting.

Job Exercise for
CHAPTERS 28–30

Purpose:

Interpret an outdoor recreation activity.

Research:

Choose an outdoor recreation activity and research jobs related to this activity.

Procedure:

Write a detailed report on your chosen outdoor recreational activity and related jobs. Your report should contain the following information:

Introduction
Explain the activity.
What jobs are available?
What training is required?
What equipment is used in the activity?
Where is the activity done?
What is the cost related to performing the activity?

Observations:

1. How are *you* involved in the activity?

2. How can others become involved in the activity?

SECTION VII
Energy, Mineral, and Metal Resources

CHAPTER 31
Fossil Fuel Management

TEST YOUR KNOWLEDGE

Complete the following:

1. The United States is a leading nation of the world because of our _____ _____.
2. Fossil fuels are minerals formed over time from compressed _____.
3. Fossil fuels include:
   a.
   b.
   c.
   d.
   e.
4. Coal is rock formed from plants in _____ areas.
5. Coal is classified according to its _____ content.
6. The lowest carbon coal is _____.
7. The second step in coal formation is _____ coal.
8. With high pressure, subbituminous coal is transformed into _____ coal.
9. The hardest and oldest coal is _____ coal.
10. Most of the coal found today is found in _____ and seams.
11. As coal burns, a poisonous gas called _____ _____ is released.
12. Scientists believe that sulfur dioxide contributes to the problem of _____ _____.
13. Coal comes from _____ mines or from _____ mines.
14. Surface mine coal is taken from a ditch called a _____.
15. The three types of underground mines include:
    a.
    b.
    c.
16. Room and pillar, and longwall, are two systems of removing coal from an _____ mine.
17. As oil was formed, it was squeezed into rock openings and _____ rocks.
18. Machines used to find oil include the gavimeter, magnetometer, and the _____.
19. The hoisting apparatus for the drilling ring is called a _____.

20. Types of off-shore operations include:

    a.

    b.

    c.

    d.

21. Crude oil is distilled into products including:

    a.

    b.

    c.

22. Fuels include:

    a.

    b.

    c.

    d.

    e.

    f.

    g.

23. Lubricants include:

    a.

    b.

    c.

    d.

24. Petrochemicals include:

    a.

    b.

    c.

    d.

    e.

    f.

    g.

25. Ways to conserve oil are:

    a.

    b.

    c.

26. The gas industry can be divided into these three areas:

    a.

    b.

    c.

27. Natural gas is the end product, but during processing, these three other gases are processed:

    a.

    b.

    c.

28. When the demand for natural gas is low, the excess can be stored in old _____ wells.

29. LNG is natural gas that has been cooled to _____ degree(s) F.

30. Most oil shale and tar sands contain only _____ percent oil.

CHAPTER 31
Fossil Fuel Management

ACTIVITY

Purpose:
Calculate miles per gallon.

Research:
1. The formula for miles per gallon is:

    $$\frac{\text{Miles}}{\text{Gallons}} = \text{Miles per gallon (round to the nearest 10th)}$$

2. The formula for miles is:

    $$\text{Gallons} \times \text{Miles per gallon} = \text{Miles (round to the nearest mile)}$$

3. The formula for gallons is:

    $$\frac{\text{Miles}}{\text{Miles per gallon}} = \text{Gallons (round to nearest 100th)}$$

Procedure:
1. Using the formula provided, calculate miles per gallon.

2. Using the formula provided, calculate miles.

3. Using the formula provided, calculate gallons.

Calculate miles per gallon

Trip	Miles	Gallons	Miles/gallon
1	100	10	
2	200	12	
3	350	20	
4	1250	58	
5	1500	100	
6	3000	120	

Calculate gallons

Trip	Miles	Gallons	Miles/gallon
1	250		12
2	1500		15
3	750		17
4	600		21
5	500		30
6	1350		38

Calculate miles

Trip	Miles	Gallons	Miles/gallon
1		9.7	18
2		11.2	21
3		27.5	22
4		20.6	25
5		106.5	27
6		142.2	32

Observations:

Show your calculations on a separate sheet.

CHAPTER 32
Alternative Energy Sources Management

TEST YOUR KNOWLEDGE

Complete the following:

1. List the main types of alternative energy sources.
    a.
    b.
    c.
    d.
    e.
    f.
    g.
    h.
    i.
2. Solar energy is more abundant, less exhaustible, and more _____-free than any other energy source.
3. Solar energy systems can be of the _____ or _____ type.
4. _____ systems capture, store, and distribute the sun's energy.
5. Passive systems rely on the natural airflow for _____.
6. Parts of an active system include:
    a.
    b.
    c.
7. Research is being done to convert sunlight to electricity by the use of _____ _____.
8. All matter is composed of _____.
9. Radium emits:
    a.
    b.
    c.
10. Rays split atomic nuclei in a process called _____.
11. During the fission process, _____ is given off.
12. A _____ _____ means that once the fission process has started, it continues on its own.
13. _____ rods can be used to control or stop the reaction.

14. The three main concerns over the use of nuclear power are:
    a.
    b.
    c.
15. _____ energy involves tapping underground reservoirs in volcanically active areas.
16. The two main disadvantages of geothermal energy are:
    a.
    b.
17. Alcohol is produced by growing yeasts in a _____ solution.
18. Yeasts take in sugar, proteins, vitamins, and minerals and give off carbon dioxide and _____.
19. Alcohol blended with gasoline makes _____.
20. Methane is produced as _____ bacteria decompose waste.
21. Methane gas can be produced artificially in a _____.
22. _____ is water power.
23. Harnessing the energy from the tides is called _____ _____.
24. Wind power requires the use of a _____ _____ like a DC battery.
25. Wood is a good source of heat energy, but it is less _____ to burn.

CHAPTER 32
Alternative Energy Sources Management

ACTIVITY

Purpose:
Calculate electrical costs for appliances.

Research:
1. A watt-hour is one watt for one hour.

2. A kilowatt-hour indicates the use of 1,000 watts of electricity for one hour.

3. The cost of residential electricity is charged by the kilowatt-hour.

$$\frac{\text{Wattage} \times \text{time in hours}}{1000} \times \text{Price per kilowatt} = \text{Electricity cost}$$

Example:

$$\frac{120 \times 2}{1000} = .24 \qquad .24 \times .10 = .024$$

Procedure:
Complete the following table to determine cost:

Appliance	Watts	Time	Price	Cost
Range	12000	1 hr.	.10	
Dishwasher	1200	3 hr.	.10	
Coffeemaker	1100	8 hr.	.10	
Washer	1200	5 hr.	.10	
Dryer	9000	5 hr.	.10	
Water heater	5000	10 hr.	.10	

CHAPTER 33
Metals and Minerals

TEST YOUR KNOWLEDGE

Complete the following:

1. Minerals are _____ resources.
2. Ferrous means _____.
3. Iron ore is the most important _____ metal.
4. Kinds of iron include:
   a.
   b.
   c.
   d.
5. Low-grade ores are called _____.
6. When iron-ore is processed, it is made into _____.
7. The _____ process is a cheap, practical method of producing steel.
8. Separating the ore from the rock can be called:
   a.
   b.
   c.
   d.
9. Smelting uses coal to remove the _____ from the ore concentrates.
10. Molten iron is poured into molds to form bars called _____ iron.
11. Pig iron bars are processed into steel:
    a.
    b.
    c.
    d.
12. When other ferrous metals are combined with iron, they are called _____.
13. Ferroalloys include:
    a.
    b.
    c.
    d.
    e.
    f.

14. Nonferrous metals include:
    a.
    b.
    c.
    d.
    e.
    f.
15. Copper:
    a. is ductile as well as _____
    b. half of all copper is used in _____ products
    c. copper mixed with tin makes _____
    d. copper mixed with zinc makes _____
    e. much of the copper that is used is _____, but even more will have to be
16. Aluminum:
    a. is one of the most common _____ on earth
    b. _____ ore contains large amounts of aluminum
    c. it takes _____ percent as much energy to produce recycled aluminum as it takes to produce it from ore
17. About 60 percent of the lead used comes from _____ sources.
18. The most common use of zinc is galvanizing for rust _____.
19. Tin is combined with copper to make _____.
20. _____ is the only metal stable in a liquid state.
21. Plant minerals include:
    a.
    b.
    c.
22. Most nitrogen used comes from _____.
23. Phosphorus is found as _____ _____.
24. Potassium occurs as _____ in beds like coal.

CHAPTER 33
Metals and Minerals

ACTIVITY

Purpose:

Perform a hardness test.

Procedure:

1. Locate a hardness test kit.

2. Locate the samples provided by the instructor.

Observations:

1. Follow the instructions in the kit to make hardness tests on the samples.

2. Complete this table:

Sample	Hardness
1	
2	
3	
4	
5	

CHAPTER 34
Careers in Energy, Mineral, and Metal Resources

TEST YOUR KNOWLEDGE

Complete the following:

1. A person employed in the oil and natural gas field can work in the following:
   a.
   b.
   c.
   d.
   e.
2. Petroleum _____ analyze and interpret information gathered by the exploration team.
3. _____ study fossil remains to locate oil-bearing layer rock.
4. _____ study the physical and chemical properties of minerals.
5. _____ determine the rock layers most likely to contain oil and natural gas.
6. _____ examine and interpret aerial photographs of land surfaces.
7. A _____ investigates the history of the formation of the earth's crust.
8. A geometry prospector manages the operation and maintenance of electronic _____ equipment.
9. The overall planning of the drilling operation is the responsibility of a _____ _____.
10. A typical drilling crew includes:
    a.
    b.
    c.
    d.
11. A gas _____ operates a unit that removes water from natural gas.
12. Skilled workers involved in the petroleum industry include:
    a.
    b.
    c.
    d.
    e.
13. Most jobs in the petroleum field are rugged, _____ jobs.
14. Starting salaries for petroleum engineers average _____.
15. About 85 percent of the people in the coal industry are _____ workers.

131

16. Mining jobs can be divided into three areas:
    a.
    b.
    c.
17. List the two methods of obtaining underground coal:
    a.
    b.
18. When earth and stone are removed above the coal seam, it is called _____ mining.
19. Miners are subject to _____ conditions.
20. Production miner workers averaged _____ per hour in 1998.
21. The nuclear energy field primarily employs:
    a.
    b.
    c.
    d.
22. There are about _____ types of jobs in the steel industry.
23. Earnings in the steel industry have been _____.

CHAPTER 34
Careers in Energy, Mineral, and Metal Resources

ACTIVITY

Purpose:

Answer interview questions.

Research:

1. Review the following information about what is needed at and what to know about interviews:

    Bring a pen
    Bring personal information including résumé
    Bring social security number
    Be prepared to discuss your specific skills
    Be positive
    Never answer with just a yes or no
    Answer all questions to the best of your ability
    Do not be afraid to say you do not know
    Answer all questions honestly
    Discuss salary range at the end of the interview
    Be prepared to ask two good questions at the end of the interview

2. Choose a job in the energy, metal, and mineral industry and research the job requirements.

Procedure:

Answer the following questions as if you were applying for the job in Research, 2:

1. What are your future plans?

2. To what organizations do you belong?

3. Why do you think you would like to work for this company?

4. Do you have any special abilities or skills?

5. Do you have any similar work experience?

6. Do you have any hobbies?

7. What do you want to be doing five years from now?

8. Why do you want this job?

9. Why do you think you are qualified for this job?

10. What are your outside interests?

11. What have you done that shows initiative and willingness to work?

Observations:
What are two questions you would ask an interviewer at the end of an interview?

Job Exercise for
CHAPTERS 31–34

Purpose:
Write a letter of application.

Research:
A letter of application should be written for every position you apply for. The first paragraph should:

    a. Tell the employer exactly what job you are applying for.

    b. Tell where you heard about the position.

    c. Include an expression of interest in the company.

The second paragraph should:

    a. Sell yourself by highlighting main accomplishments.

    b. Provide your current employment information. This should be kept brief because this information will be in your résumé.

    c. List important educational information.

The third paragraph should:

    a. Note whether you have enclosed supplemental information.

    b. Include your telephone number and when it is best to contact you.

    c. Reexpress your interest in the job.

    d. Thank the employer for their time.

Example

```
Your Street Address
Your City, State, Zip
Date

Name of person
Title of person
Name of Company
Street Address
City, State, Zip

Dear _____:

I am interested in working for your department.  I learned from my
employment counselor that you are expanding services.  I understand that
your company is environmentally conscious, and I would like to become an
employee of such a company.

I have taken natural resources classes to support my interest in your
industry.  I have been involved in our local "Adopt A Park" program,
donating over 250 hours of volunteer service.  I am a member of our local
FFA Chapter, and was on the state FFA Forestry Contest team this past year.
Although I do not have full-time work experience, I feel I have the
enthusiasm and recent academic preparation that would foster a successful
employment relationship with your company.

Please find enclosed a personal resume containing further information.
I am excited about the possibility of working for your company.  I look
forward to hearing from you at your convenience.  You may contact me at
555-1234.

Thank you for your consideration and time.

Sincerely yours,


Your Name
```

Procedure:

Write a letter of application for one of the following jobs:

petroleum engineer	steel industry salesperson
dragline operator	miner
engineering technician	drilling crew laborer
steel industry laborer	gas treater
soil conservation technician	oceanographer
geologist	forestry technician
game biologist	fishery biologist
fish and wildlife technician	

SECTION VIII
Advanced Concepts

CHAPTER 35
Advanced Concepts in Natural Resources Management

TEST YOUR KNOWLEDGE

Complete the following:

1. _____ implies doing something.
2. _____ an area in its natural state is an intentional management practice.
3. The first step in management is a _____.
4. Selecting the _____ management practice from among alternative solutions is a complex process.
5. Our natural resources are held in _____ by the government for the benefit of the people.
6. An important change in natural resources management has been that the public _____ in the decision-making process.
7. _____ advocate the management of nature while using the resources to benefit people.
8. _____ change some aspect of nature to make it more valuable and profitable.
9. _____ take something from nature and move it for a profit.
10. The _____ we make as a society must consider our wants and needs while considering that our resources need protection, maintenance, and enhancement.
11. Resource _____ is best assigned according to the use of the resources.
12. Water, soil, and air are considered _____ resources.
13. The goal of a _____ yield management is to protect the quantity and quality of the resource.
14. Management for a sustained yield means the use of renewable resources in such a way as to allow a constant rate of _____ indefinitely.
15. The three E's of resources management are:
    a.
    b.
    c.
16. Resources are not equally _____ around the world.
17. Common _____ are in theory owned by everyone but in reality owned by no one.
18. Whether the resources are owned privately or not has major implications for _____.
19. Everything we do with natural resources is done with the interests of _____ at heart.
20. _____ population is the number one environmental problem.

CHAPTER 35
Advanced Concepts in Natural Resources Management

ACTIVITY

Purpose:

Determine direction using a compass.

Materials Needed:

compass
100' tape

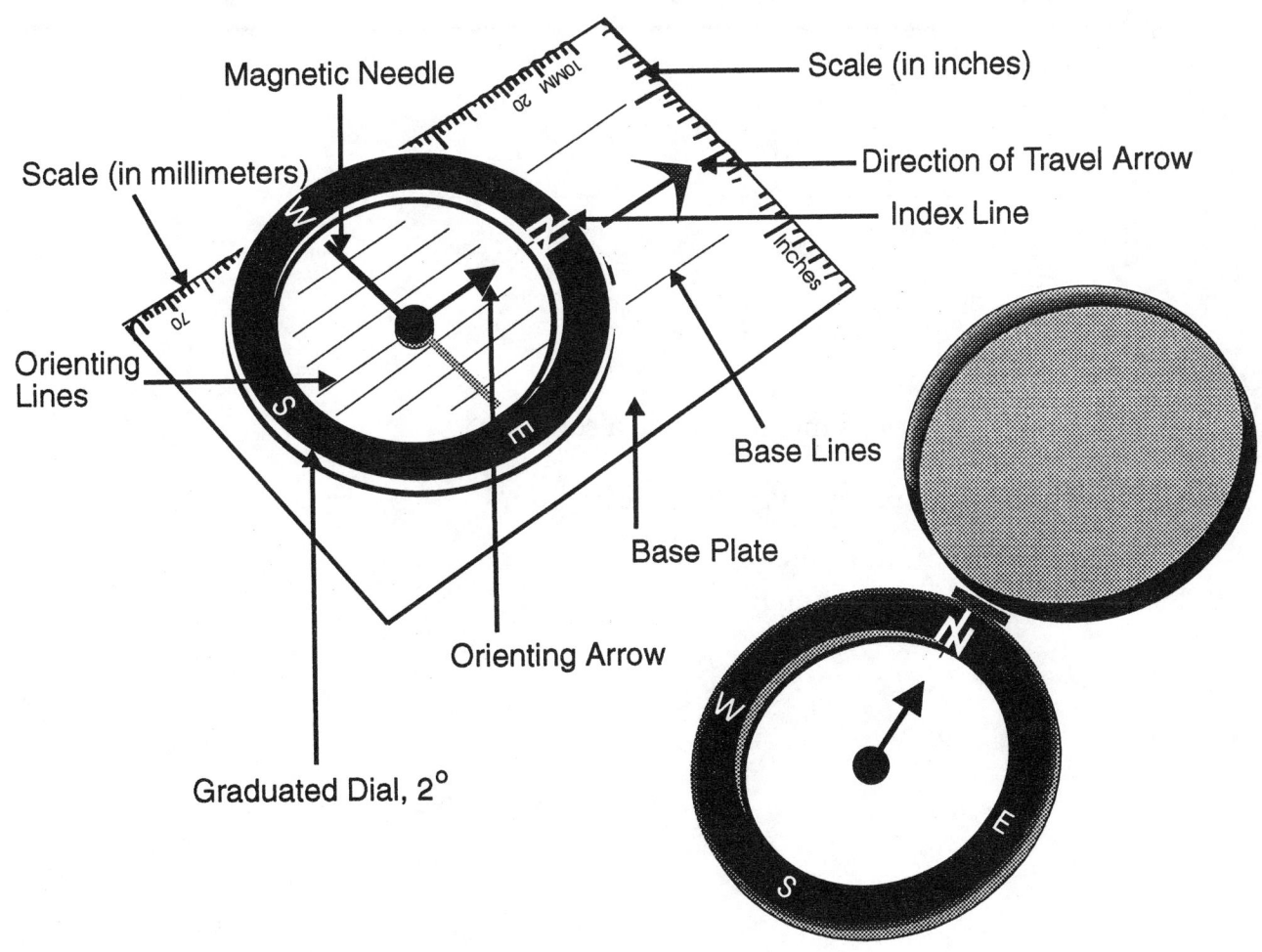

Procedure:

1. Locate the site markers and starting point prepared by the instructor.

2. Stand at Site 1.

3. Hold the compass in your hand in the direction of Site 2.

4. Rotate the compass dial until the orienting arrow overlaps the floating magnetic needle.

5. Read the bearing in degrees from the dial.

6. Measure the distance with the tape.

7. Move to the next point.

8. Repeat the procedure for all site markers.

Site	Bearing	Distance
1		
2		
3		
4		
5		

Observations:

1. In what applications might you use a compass?

2. What groups of people might use a compass?

Job Exercise for
CHAPTER 35

Purpose:

Complete a résumé.

Research:

A résumé is an advertisement of facts about yourself and your skills. It should contain accurate facts. Your résumé should be on 8½" by 11" paper and should be no more than two pages. You should include personal information, educational information, work-related information, and references. Your name should appear prominently at the top of the résumé.

Procedure:

Obtain résumés from people you know. Study the various forms. Choose a format that best suits you. One example is listed on the following page.

Example

	Your Name
ADDRESS:	1400 John David Drive Anytown, SD 57001 (605) 555-1234 Messages taken at: (605) 555-1141 MyE-mailAddress@aol.com
PROFESSIONAL OBJECTIVE:	Seeking a position as an assistant manager to retail outlet in the south side of the city.
EDUCATION:	South Dakota State University, Brookings, SD, 1998-01. Bachelors Degree in Business Management. Academic preparation in communication, business management, and supervision
WORK EXPERIENCE:	2001. Helland Stores Inc., Anytown, S.D. Assisted the manager with inventory and ordering. 1998-2001. Mytown Video, Mytown, S.D. Cashier.
ORGANIZATIONS:	Ducks Unlimited, Recruitment committee chairman.
REFERENCES:	Mike L. Jones, Helland Stores Inc. (605) 555-1124 Dale E. Everson, Mytown Video. (605) 555-1243 R.L. Jones, Mytown Express. (605) 555-4321

Please Note: Next to References you may also say "Available upon request." If you do choose to list your references, make sure that you have the person's permission first.

Observation:

Write your résumé.